GRAVITY

GEORGE GAMOW

Illustrations by the Author

DOVER PUBLICATIONS, INC.
Mineola, New York

Bibliographical Note

This Dover edition, first published in 2002, is an unabridged republication of the work originally published (as part of the Science Study Series) by Anchor Books, Doubleday & Company, Inc., Garden City, New York, in 1962. A new Preface has been specially prepared for this edition.

Library of Congress Cataloging-in-Publication Data

Gamow, George, 1904-1968.
    Gravity / George Gamow ; illustrated by the author.
        p. cm.
    Originally published: Garden City, N.Y. : Anchor Books, 1962, in series: Science study series.
    Includes bibliographical references and index.
    ISBN 0-486-42563-0 (pbk.)
        1. Gravitation. I. Title.

QC178 .G3 2002
531'.14—dc21

2002034827

Manufactured in the United States of America
Dover Publications, Inc., 31 East 2nd Street, Mineola, N.Y. 11501

GRAVITY

BIOGRAPHY

In 1954 Dr. George Gamow, who had been specializing in the application of nuclear physics to problems in astrophysics and cosmology, made a mathematical suggestion about cell chemistry and thereby set a pattern for DNA research in genetics that has turned out to be quite valuable. This episode, in the telling of which one must mention nuclear physics, astrophysics, cosmology, mathematics, chemistry, and biology, is illustrative of Dr. Gamow's many-faceted career in science but fails to do his versatility full justice. Literary critics in both the United States and England have acclaimed him the best living interpreter of science to the layman, have delighted in his scientific fantasies and discovered poetry in his writing. To secure a reputation in either science or literature ordinarily is accounted achievement enough for one man, but Dr. Gamow has not been satisfied with both. As readers of this little book will be privileged to see, Dr. Gamow also has tackled the fine arts, democratically drawing inspiration from both the Sunday comics and the masterpieces of Sandro Botticelli. (If at first glance you do not detect the Botticelli influence in Dr. Gamow's portrait of the late Albert Einstein on page 119, at least you have assurance from the artist himself that it is there.) And in all these separate fields of creativity, Dr. Gamow has been about equally prolific.

Dr. Gamow was born on March 4, 1904, in Odessa, Russia. In early youth he turned to science and spent a year studying paleontology. This experience, he said later, equipped him "to tell a dinosaur from a cat by the shape of the little toes." He entered the University of Leningrad, from which he received a Ph.D. degree in 1928, and spent a year at the University of Göttingen, in Germany, on a traveling fellowship. In 1928-29 he worked with Niels Bohr in Copenhagen and in 1929-30 with Ernest Rutherford at the Cavendish Laboratory, Cambridge, England.

Dr. Gamow was twenty-four when he made his first major contribution to physical theory. Concurrently, but independently, he, on the one hand, and the American physicist E. U. Condon and the British physicist R. W. Gurney, on the other, explained the emission of alpha-particles from radioactive atoms by applying to the process the then new methods of wave mechanics. Two years later, in 1930, he made the successful prediction that protons would be more useful than alpha-particles in the experiments popularly known as "atom-smashing," and in the same year he suggested the liquid drop model for the nuclei of heavy elements. In 1929 he collaborated with R. Atkinson and F. Houtermans in formulating the theory that the sun's heat and light resulted from thermonuclear processes, and his theory of the origin of chemical elements through neutron capture dominated cosmological thinking at one period in the 1940s. His contribution to DNA theory was the suggestion that the four nucleotides of the DNA molecule compose a code whose different combinations act as templates in the organization of the various amino acid molecules.

Dr. Gamow's personal characteristics are almost as formidable as his creative achievements. A giant, six feet three and well over 225 pounds, he is given to puckish humor, as readers of his *Mr. Tompkins* fantasies well know. When he and his student, R. Alpher, signed their names to the preliminary calculations of their paper, *The Origin of Chemical Elements,* in 1948, Gamow commented, "Something is missing," and, crediting Hans Bethe *in absentia,* made the signature "Alpher, Bethe and Gamow." He speaks six languages and is a frequent and popular lecturer with a heavily accented delivery that moved a friend to observe that the six languages were all different dialects of one language—"Gamovian."

His ability as a linguist, however accented, reflects the ground he has covered in his professional career. After his studies with Bohr and Rutherford, he returned to Russia as Master in Research at the Academy of Sciences in Leningrad but left his native land for good in 1933. He lectured in Paris and London, at the University of Michigan summer school, and then joined the faculty of George Washington University, Washington, D.C., where he was professor of physics from 1934 to 1956. He became a United States citizen in 1940 and acted as a Navy, Army, Air Force and Atomic Energy Commission consultant during and after World War II. Since 1956 he has been on the faculty of the University of Colorado, Boulder.

Dr. Gamow has written many technical papers and one technical book, *Atomic Nucleus* (Oxford University Press, 1931, revised 1937 and 1949). His popular writing includes numerous *Scientific American* articles and the following books:

Mr. Tompkins in Wonderland, Cambridge University Press, 1939

Mr. Tompkins Explores the Atom, Cambridge University Press, 1943

Mr. Tompkins Learns the Facts of Life, Cambridge University Press, 1953

Atomic Energy in Cosmic and Human Life, Cambridge University Press, 1945

The Birth and Death of the Sun, Viking Press, 1941

Biography of the Earth, Viking Press, 1943

One, Two, Three . . . Infinity, Viking Press, 1947

Creation of the Universe, Viking Press, 1952

Puzzle-Math (with M. Stern), Viking Press, 1958

The Moon, H. Schuman, 1953

Matter, Earth and Sky, Prentice Hall, 1958

Physics: Foundation and Frontiers (with J. Cleveland), Prentice Hall, 1960

Atom and Its Nucleus, Prentice Hall, 1961

Biography of Physics, Harper and Brothers, 1961

He took up illustrating for the second *Mr. Tompkins* book when World War II interrupted communication between him and the English artist who had worked with him on the earlier book of the series. In 1956 he received the Kalinga Prize from UNESCO for his popular interpretations of science for lay readers.

Dr. Gamow was a member of the Academy of Science of the U.S.S.R. until, as he says, he was "fired after leaving Russia." He is a member of the Royal Danish Academy of Sciences and the National Academy of Sciences of the United States.

PREFACE TO THE DOVER EDITION

George Gamow (pronounced *Gam*-off) was a first-rate scientist who is credited with major advances in nuclear physics and cosmogenesis. But, as the preceding biographical note indicates, he was equally well known, at the time of his death in 1968, for his lucid, lively interpretations of science for the layperson. According to a grateful reporter who sought Gamow's help in understanding the physics behind the Bikini H-bomb test in 1949, "He can always take the most technical information and make it simple." Another journalist at that cataclysmic event declared that Gamow was "the only scientist in America with a real sense of humor." Both of these traits come through in Gamow's writing, and a surprising number of scientists—including more than one Nobel laureate—have credited his books with having stimulated their childhood interest in science. Gamow delighted readers with his series of insidiously informative books featuring the timid but curious bank clerk C. G. H. Tompkins, with his *Scientific American* articles, and with classics such as *One, Two, Three . . . Infinity*, *The Birth and Death of the Sun*, and the little book you are now holding in your hands.

As a youngster in Odessa, Gyorgi Antonovich Gamow became fascinated by astronomy and examined the sky assiduously through the little telescope his father had given him for his thirteenth birthday. Later at school he gave himself over to the study of mathematics and physics. In 1928, following six years of study at the University of Leningrad (and one at Novorossia University in Odessa), he attended summer school in Göttingen, Germany, which at that time was feverishly buzzing with the recently developed quantum theory. Through his research, he was able to explain nuclear radioactivity as a quantum-mechanical phenomenon— an explanation that confirmed the experimental findings of Ernest Rutherford and other cutting-edge scientists at Cambridge University's Cavendish Laboratory. On the basis of this research, Gamow received his doctorate from the University of Leningrad and an invitation from Niels Bohr to join him for a year at the Theoretical Physics Institute of the University of Copenhagen. In Copenhagen, Gamow proposed the "liquid drop" model for the nuclei of heavy elements—which said, essentially, that for theoretical purposes scientists could treat atomic nuclei as droplets of fluid. He also collaborated with the Austrian physicist Fritz Houtermans and the English astrophysicist Robert Atkinson in calculating the rate of thermonuclear reactions in the Sun and other stars.

In 1931, the year of his marriage to Lyubov Vokhminzeva (known as Rho), Gamow became a professor at the University of Leningrad. But he and his wife were aching to get away from the strictures of life in the Soviet Union, and after an unsuccessful attempt (in 1932) to escape by paddling a kayak 170 miles across the Black Sea to Turkey, they got their real chance when the government appointed him to represent the Soviet Union at a theoretical physics conference in Brussels in October 1933. After the conference, Gamow accepted an invitation to lecture at the University of Michigan, and in August 1934 he and his wife moved to Washington, D.C., where for the next twenty-two years he was the shining star of George Washington University's physics department. In 1956 he got a divorce from Rho and moved west to teach at the University of Colorado, where he remained for the rest of his life.

Not long after coming to Washington, Gamow shifted his gaze to the stars and applied his knowledge of nuclear physics to questions of stellar birth, evolution, and energy generation—focusing particularly on the internal structure of red giant stars. But the contribution for which he is probably best remembered is his development of the Big Bang model[1] of the universe, including his speculations about the creation of the elements and his prediction (later empirically verified) of the existence of cosmic background radiation. Edwin Hubble in 1929 had found observational evidence of the universe's expansion, which seemed to corroborate the idea—put forth by a Russian mathematician and a Belgian astronomer—that the universe, rather than being eternal, might have had a beginning. Gamow, his student Ralph Alpher, and their colleague Robert Herman developed the physical aspects of this idea, notably in Gamow and Alpher's[2] famous 1948 paper, "The Origin of Chemical Elements." They proposed that the universe originated from a single point—a superdense black hole or singularity that contained all of the matter in the universe in the form of neutrons with a temperature of ten billion degrees. Eight to twenty billion years ago, this entity underwent an explosion of unimaginable magnitude, spewing matter in every

---

[1] The name, meant to be disparaging, was coined by the British astronomer Fred Hoyle, who advocated an opposing steady-state model that now seems to have been a dead end.

[2] As previously mentioned in the biographical note, Gamow could not resist whimsically adding Hans Bethe's name as a co-author.

direction. Such an event, Gamow predicted, would have left throughout the modern cosmos a background microwave radiation with a temperature of five degrees K. It would also have provided the conditions necessary for the formation of the elements, most of which must have been formed, through fusion, within the first few minutes of the explosion. The heavier elements were formed later, within the crucible of the stars.

Gamow's intellect was always too restless to be pigeonholed. In 1953, as discussed in the preceding biographical note, he astonished many of his colleagues by leaping precipitously into a field he would seem to know nothing about: genetic coding. After reading Francis Crick and James Watson's famous paper describing the structure of DNA, Gamow sent Crick a letter positing a mathematical relationship between that structure and the amino acid sequence of proteins. In the end, that relationship did not pan out. But the suggestion was so brilliantly perceptive that it helped to stimulate huge advances in the understanding of genetic coding, and Crick made special note of it in his 1962 speech accepting the Nobel Prize: "It is one of the more striking generalizations of biochemistry . . . that the twenty amino acids and the four bases, are, with minor reservations, the same throughout Nature. As far as I am aware the presently accepted set of twenty amino acids was first drawn up by Watson and myself in the summer of 1953 in response to a letter of Gamow's."

"Gamow was fantastic in his ideas," Edward Teller once said. "More often wrong than right. Always interesting . . . and when his idea was not wrong it was not only right, it was new."

T. N. R. Rogers

To
QUIGG NEWTON
who reads all my books

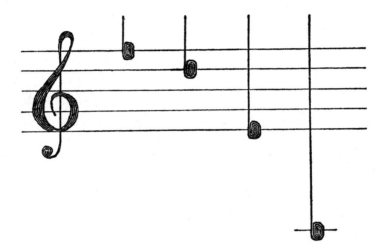

PREFACE

Gravity rules the Universe. It holds together the one hundred billion stars of our Milky Way; it makes the Earth revolve around the Sun and the Moon around the Earth; it makes ripened apples and disabled airplanes fall to the ground. There are three great names in the history of man's understanding of gravity: Galileo Galilei, who was the first to study in detail the process of free and restricted fall; Isaac Newton, who first had the idea of gravity as a universal force; and Albert Einstein, who said that gravity is nothing but the curvature of the four-dimensional space-time continuum.

In this book we shall go through all three stages of the development, devoting one chapter to Galileo's pioneering work, six chapters to Newton's ideas and their subsequent development, one chapter to Einstein, and one chapter to post-Einsteinian speculations concerning the relation between gravity and other physical phenomena. The emphasis on the "classics" in this outline grows from the fact that the theory of universal gravity *is* a classical theory. It is very probable that there is a hidden relation between gravity on the one hand and the electromagnetic field and material particles on the other, but nobody is prepared today to say what kind of relation it is. And there is no way of foretelling how soon any further important progress will be made in this direction.

Considering the "classical" part of the theory of gravitation, the author had to make an important decision about the use of mathematics. When Newton first conceived the idea of Universal Gravity, mathematics was not yet developed to a degree that could permit him to follow all the astronomical consequences of his ideas. Thus Newton had to develop his own mathematical system, now known as the differential and integral calculus, largely in order to answer the problems raised by his theory of universal gravitation. Therefore it seems reasonable, and not only from the historical point of view, to include in this book a discussion of the elementary principles of calculus, a decision which accounts for a rather large number of mathematical formulas in the third chapter. The reader who has the grit to concentrate on that chapter will certainly profit by it as a basis for his further study of physics. On the other hand, those who are frightened by mathematical formulas can skip that chapter without much damage to a general understanding of the subject. But if you want to learn physics, *please do try* to understand Chapter 3!

George Gamow

University of Colorado
January 13, 1961

CONTENTS

GRAVITY

Chapter 1

HOW THINGS FALL

The notion of "up" and "down" dates back to time immemorial, and the statement that "everything that goes up must come down" could have been coined by a Neanderthal man. In olden times, when it was believed that the world was flat, "up" was the direction to Heaven, the abode of the gods, while "down" was the direction to the Underworld. Everything which was not divine had a natural tendency to fall down, and a fallen angel from Heaven above would inevitably finish in Hell below. And, although great astronomers of ancient Greece, like Eratosthenes and Aristarchus, presented the most persuasive arguments that the Earth was round, the notion of absolute up-and-down directions in space persisted through the Middle Ages and was used to ridicule the idea that the Earth could be spherical. Indeed, it was argued that if the Earth were round, then the antipodes, the people living on the opposite side of the globe, would fall off the Earth into empty space below, and, far worse, all ocean water would pour off the Earth in the same direction.

When the sphericity of the Earth was finally established in the eyes of everyone by Magellan's round-the-world trip, the notion of up-and-down as an absolute direction in space had to be modified. The terrestrial globe was considered to be resting at the center of the

Universe while all the celestial bodies, being attached to crystal spheres, circled around it. This concept of the Universe, or cosmology, stemmed from the Greek astronomer Ptolemy and the philosopher Aristotle. The natural motion of all material objects was toward the center of the Earth, and only Fire, which had something divine in it, defied the rule, shooting upward from burning logs. For centuries Aristotelian philosophy and scholasticism dominated human thought. Scientific questions were answered by dialectic arguments (i.e., by just talking), and no attempt was made to check, by direct experiment, the correctness of the statements made. For example, it was believed that heavy bodies fall faster than light ones, but we have no record from those days of an attempt to study the motion of falling bodies. The philosophers' excuse was that free fall was too fast to be followed by the human eye.

The first truly scientific approach to the question of how things fall was made by the famous Italian scientist Galileo Galilei (1564–1642) at the time when science and art began to stir from their dark sleep of the Middle Ages. According to the story, which is colorful but probably not true, it all started one day when young Galileo was attending a Mass in the Cathedral of Pisa, and absent-mindedly watched a candelabrum swinging to and fro after an attendant had pulled it to the side to light the candles (Fig. 1). Galileo noticed that although the successive swings became smaller and smaller as the candelabrum came to rest, the time of each swing (oscillation period) remained the same. Returning home, he decided to check this casual observation by using a stone suspended on a string and measuring the swing period by counting his pulse. Yes, he was right; the period remained almost the same while the swings became shorter and shorter. Being of an inquisitive turn of mind, Galileo started a series of

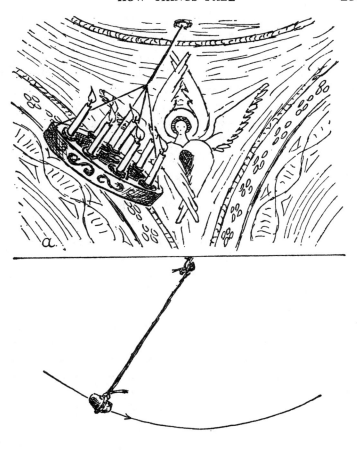

Fig. 1. A candelabrum (a) and a stone on a rope (b) swing with the same period if the suspensions are equally long.

experiments, using stones of different weights and strings of different lengths. These studies led him to an astonishing discovery. Although the swing period depended on the string's length (being longer for longer strings), it was quite independent of the *weight* of the suspended stone. This observation was definitely contradictory to the accepted dogma that heavy bodies fall faster than light ones. Indeed, the motion of a pendulum is nothing but the free fall of a weight deflected from a vertical direction by a restriction imposed by a string, which makes the weight move along an arc of a circle with the center in the suspension point (Fig. 1).

If light and heavy objects suspended on strings of equal length and deflected by the same angle take equal time to come down, then they should also take equal time to come down if dropped simultaneously from the same height. To prove this fact to the adherents of the Aristotelian school, Galileo climbed the Leaning Tower of Pisa or some other tower (or perhaps deputized a pupil to do it) and dropped two weights, a light and a heavy one, which hit the ground at the same time, to the great astonishment of his opponents (Fig. 2).

There seems to be no official record concerning this demonstration, but the fact is that Galileo was the man who discovered that the velocity of free fall does not depend on the mass of the falling body. This statement was later proved by numerous, much more exact experiments, and, 272 years after Galileo's death, was used by Albert Einstein as the foundation of his relativistic theory of gravity, to be discussed later in this book.

It is easy to repeat Galileo's experiment without visiting Pisa. Just take a coin and a small piece of paper and drop them simultaneously from the same height to the floor. The coin will go down fast, while the piece of paper will linger in the air for a much longer period of time. But if you crumple the piece of paper and roll it into a little ball, it will fall almost as fast as the coin. If you had a long glass cylinder evacuated of air, you would see that a coin, an uncrumpled piece of paper,

Fig. 2. Galileo's experiment in Pisa.

and a feather would fall inside the cylinder at exactly the same speed.

The next step taken by Galileo in the study of falling bodies was to find a mathematical relation between the time taken by the fall and the distance covered. Since the free fall is indeed too fast to be observed in detail by the human eye, and since Galileo did not possess such modern devices as fast movie cameras, he decided to "dilute" the force of gravity by letting balls made of different materials roll down an inclined plane instead of falling straight down. He argued correctly that, since the inclined plane provides a partial support to heavy objects placed on it, the ensuing motion should be similar to free fall except that the time scale would be lengthened by a factor depending on the slope. To measure time he used a water clock, a device with a spigot that could be turned on and off. He could measure intervals of time by weighing the amounts of water that poured out the spigot in different intervals. Galileo marked the successive position of the objects rolling down an inclined plane at equal intervals of time.

You will not find it difficult to repeat Galileo's experiment and check on the results he obtained.* Take a smooth board 6 feet long and lift one end of it 2 inches from the floor, placing under it a couple of books (Fig. 3a). The slope of the board will be $\dfrac{2}{6 \times 12} = \dfrac{1}{36}$, and this will also be the factor by which the gravity force acting on the object will be reduced. Now take a metal cylinder (which is less likely to roll off the board than a ball) and let it go, without pushing, from the top end of the board. Listen to a ticking clock or a metronome (such as music students use) and mark the position of the rolling cylinder at the end of the first, second, third, and fourth seconds. (The experiment should be repeated several times to get these positions exact.) Under these conditions, consecutive distances from the top end will be 0.53, 2.14, 4.82, 8.5, and 13.0

---

* Not being an experimentalist, the author is not able to say, on the basis of his own experience, how easy it is to do Galileo's experiment. He has heard from various sources, however, that this is, in fact, not so easy and would recommend that the readers of this book try their skill at it.

inches. We notice, as Galileo did, that distances at the
end of the second, third, and fourth seconds are respectively 4, 9, 16, and 25 times the distance at the end of
the first second. This experiment proves that the veloc-

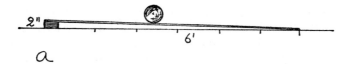

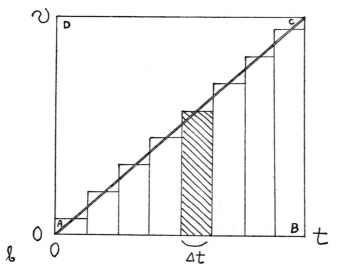

Fig. 3. (a) A rolling cylinder on an inclined plane; (b)
Galileo's method of integration.

ity of free fall increases in such a way that *the distances covered by a moving object increase as the
squares of the time of travel.* $(4 = 2^2; 9 = 3^2; 16 =
4^2; 25 = 5^2)$ Repeat the experiment with a wooden
cylinder, and a still lighter cylinder made of balsa wood,
and you will find that the speed of travel and the dis-

tances covered at the end of consecutive time intervals
remain the same.

The problem that then faced Galileo was to find the
law of change of velocity with time, which would lead
to the distance-time dependence stated above. In his
book *Dialogue Concerning Two New Sciences* Galileo
wrote that the distances covered would increase as the
squares of time if the velocity of motion was proportional to the first power of time. In Fig. 3b we give a
somewhat modernized form of Galileo's argument. Consider a diagram in which the velocity of motion v is
plotted against time t. If v is directly proportional to t
we will obtain a straight line running from $(o;o)$ to
$(t;v)$. Let us now divide the time interval from o to t
into a large number of very short time intervals and
draw vertical lines as shown in the figure, thus forming
a large number of thin tall rectangles. We now can
replace the smooth slope corresponding to the continuous motion of the object with a kind of staircase representing a jerky motion in which the velocity abruptly
changes by small increments and remains constant for a
short time until the next jerk takes place. If we make
the time intervals shorter and shorter and their number
larger and larger, the difference between the smooth
slope and the staircase will become less and less noticeable and will disappear when the number of divisions
becomes infinitely large.

During each short time interval the motion is
assumed to proceed with a constant velocity corresponding to that time, and the distance covered is equal
to this velocity multiplied by the time interval. But since
the velocity is equal to the height of the thin rectangle,
and the time interval to its base, this product is equal
to the *area* of the rectangle.

Repeating the same argument for each thin rectangle,
we come to the conclusion that the total distance

covered during the time interval (o,t) is equal to the area of the staircase or, in the limit, to the area of the triangle ABC. But this area is one-half of the rectangle $ABCD$ which, in its turn, is equal to the product of its base t by its height v. Thus, we can write for the distance covered during the time t:

$$s = \frac{1}{2} vt$$

where v is the velocity at the time t. But, according to our assumption, v is proportional to t so that:

$$v = at$$

where a is a constant known as *acceleration* or the rate of change of velocity. Combining the two formulas, we obtain:

$$s = \frac{1}{2} at^2$$

which proves that distance covered increases as the square of time.

The method of dividing a given geometrical figure into a large number of small parts and considering what happens when the number of these parts becomes infinitely large and their size infinitely small, was used in the third century B.C. by the Greek mathematician Archimedes in his derivation of the volume of a cone and other geometrical bodies. But Galileo was the first to apply the method to mechanical phenomena, thus laying the foundation for the discipline which later, in the hands of Newton, grew into one of the most important branches of the mathematical sciences.

Another important contribution of Galileo to the young science of mechanics was the discovery of the principle of *superposition of motion*. If we should throw a stone in a horizontal direction, and if there were no gravity, the stone would move along a straight line as a ball does on a billiard table. If, on the other hand, we just dropped the stone, it would fall vertically with the increasing velocity we have described. Actually, we have the superposition of two motions: the stone moves horizontally with constant velocity and at the same time falls in an accelerated way. The situation is repre-

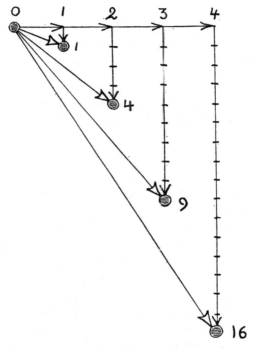

Fig. 4. A combination of horizontal motion with constant velocity, and vertical, uniformly accelerated motion.

sented graphically in Fig. 4 where the numbered horizontal and vertical arrows represent distances covered in the two kinds of motion. The resulting positions of the stone can also be given by the single (white-headed) arrows, which become longer and longer and turn around the point of origin.

Arrows like these that show the consecutive positions of a moving object in respect to the point of origin are called *displacement vectors* and are characterized by their *length* and their *direction* in space. If the object undergoes several successive displacements, each being described by the corresponding displacement vector, the final position can be described by a single displacement vector called *the sum* of the original displacement vectors. You just draw each succeeding arrow beginning from the end of the previous one (Fig. 4), and connect the end of the last arrow with the beginning of the first by a straight line. In plain words for a trivial example, a plane which flew from New York City to Chicago, from Chicago to Denver, and from Denver to Dallas, could have gone from New York City to Dallas by flying a straight course between the two cities. An alternative way of adding two vectors is to draw both arrows from the same point, complete the parallelogram and draw its diagonal as shown in Fig. 5a and b. Comparing the two drawings, one is easily persuaded that they both lead to the same result.

The notion of displacement vectors and their additions can be extended to other mechanical quantities which have a certain direction in space. Imagine an aircraft carrier making so many knots on a north-northwest course, and a sailor running across its deck from starboard to port at the speed of so many feet per minute. Both motions can be represented by arrows pointing in the direction of motion and having lengths proportional to the corresponding velocities (which must of course be expressed in the same units). What is the velocity of the sailor in respect to water? All we have to do is to add the two velocity vectors according to the rules, i.e., by constructing the diagonal of a parallelogram defined by the two original vectors.

Forces, too, can be represented by vectors showing

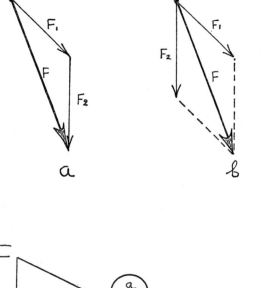

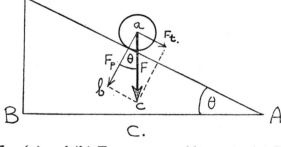

Fig. 5. (a) and (b) Two ways to add vectors; (c) Forces acting on a cylinder placed on an inclined plane.

the direction of the acting force and the amount of effort applied, and can be added according to the same rule. Let us consider, for example, the vector of the gravity force acting on an object placed on an inclined plane (Fig. 5c). This vector is directed vertically down, of course, but reversing the method of adding vectors, we can represent it as the sum of two (or more) vectors pointing in given directions. In our example we want one component to point in the direction of the inclined plane and another perpendicular to it as shown in the figure. We notice that the rectangular triangles ABC (geometry of the inclined plane), and abc (formed by vectors F, F_p and F_t) are similar, having equal angles at A and a respectively. It follows from Euclidian geometry that

$$\frac{F_2}{F} = \frac{BC}{AC}$$

and this equation justifies the statement we made concerning Galileo's experiment with the inclined plane.

Using the data obtained by experiments with inclined planes, one can find that the acceleration of free fall is $386.2 \frac{inches}{sec^2}$ (you probably are familiar with the equivalent expression "32.2 feet per second per second") or in the metric system, $981 \frac{cm}{sec^2}$. This value varies slightly with the latitude on the Earth's surface, and the altitude above sea level.

Chapter 2

THE APPLE AND THE MOON

The story that Isaac Newton discovered the Law of Universal Gravity by watching an apple fall from a tree (Fig. 6) may or may not be as legendary as the stories about Galileo's watching the candelabrum in the Cathedral of Pisa or dropping weights from the Leaning Tower, but it enhances the role of apples in legend and history. Newton's apple rightfully has a place with the apple of Eve, which resulted in the expulsion from Paradise, the apple of Paris, which started the Trojan War, and the apple of William Tell, which figured in the formation of one of the world's most stable and peace-loving countries. There is no doubt that when the twenty-three-year-old Isaac was contemplating the nature of gravity, he had ample opportunity to observe falling apples; at the time he was staying on a farm in Lincolnshire to avoid the Great Plague, which descended on London in 1665 and led to the temporary closing of Cambridge University. In his writings Newton remarks: "During this year I began to think of gravity extending to the orb of the Moon, and compared the force requisite to keep the Moon in her orb with the forces of gravity at the surface of the Earth." His arguments concerning this subject, given later in his book *Mathematical Principles of Natural Philosophy*, run roughly as follows: If, standing on the top of a moun-

Fig. 6. Isaac Newton on the Lincolnshire farm.

tain we shoot a bullet in a horizontal direction, its motion will consist of two components: a) horizontal motion with the original muzzle velocity; b) an accelerated free fall under the action of gravity force. As a result of superposition of these two motions, the bullet will describe a parabolic trajectory and hit the ground some distance away. If the Earth were flat, the bullet would always hit the Earth even though the impact might be very far away from the gun. But since the Earth is round, its surface continuously curves under the bullet's path, and, at a certain limiting velocity, the bullet's curving trajectory will follow the curvature of the globe. Thus, if there were no air resistance, the bullet would never fall to the ground but would continue circling the Earth at a constant altitude. This was the first theory of an artificial satellite, and Newton illustrated it with a drawing very similar to those we see today in popular articles on rockets and satellites. Of course, the satellites are not shot from the tops of mountains, but are first lifted almost vertically beyond the limit of the terrestrial atmosphere, and then given the necessary horizontal velocity for circular motion. Considering the motion of the Moon as a continuous fall which all the time misses the Earth, Newton could calculate the force of gravity acting on the Moon's material. This calculation, in somewhat modernized form, runs as follows:

Consider the Moon moving along a circular orbit around the Earth (Fig. 7). Its position at a certain moment is M, and its velocity perpendicular to the radius of the orbit is v. If the Moon were not attracted by the Earth, it would move along a straight line, and, after a short time interval, Δt, would be in position M' with $\overline{MM'} = v\Delta t$. But there is another component of the Moon's motion; namely, the free fall toward the Earth. Thus, its trajectory curves and, instead of arriv-

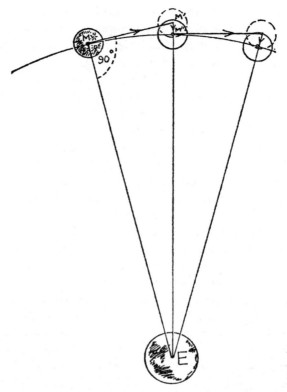

Fig. 7. Calculation of the acceleration of the Moon.

ing at M', it arrives at the point M'' on its circular orbit, and the stretch $\overline{M'M''}$ is the distance it has fallen toward the Earth during the time interval Δt. Now, consider the right triangle EMM' and apply to it the Pythagorean theorem, which says that in a right triangle the square on the side opposite the right angle is equal to the sum of the squares on the other two sides. We obtain

$$(\overline{EM''} + \overline{M''M'})^2 = \overline{EM}^2 + \overline{MM'}^2$$

or, opening the parenthesis:

$$\overline{EM''}^2 + 2\overline{EM''} \cdot \overline{M''M'} + \overline{M''M'}^2 = \overline{EM}^2 + \overline{MM'}^2$$

Since $\overline{EM''} = \overline{EM}$, we cancel these terms on both sides of the equation, and, dividing by $2\overline{EM}$ we obtain:

$$\overline{M''M'} + \frac{\overline{M''M'}^2}{2\overline{EM}} = \frac{\overline{MM'}^2}{2\overline{EM}}$$

Now comes an important argument. If we consider shorter and shorter time intervals, $\overline{M''M'}$ becomes correspondingly smaller, and both terms on the left side come closer and closer to zero. But, since the second term contains the *square* of $\overline{M''M'}$ it goes to zero faster than the first; in fact, if $\overline{M''M'}$ takes the values:

$$\frac{1}{10} ; \quad \frac{1}{100} ; \quad \frac{1}{1000} ; \quad \text{etc.}$$

its square becomes:

$$\frac{1}{100} ; \quad \frac{1}{10,000} ; \quad \frac{1}{1,000,000} ; \quad \text{etc.}$$

Thus, for sufficiently small time intervals we may neglect the second term on the left as compared with the first, and write:

$$\overline{M''M'} = \frac{\overline{MM'}^2}{2\overline{EM}}$$

which will be exactly correct, of course, only when $\overline{M''M'}$ is infinitesimally small.

Since $\overline{MM'} = v\Delta t$ and $\overline{EM} = R$, we can rewrite the above as

$$\overline{M''M'} = \frac{1}{2}\left(\frac{v^2}{R}\right)\Delta t^2$$

In discussing Galileo's studies of the law of fall, we have seen that the distance traveled during the time interval Δt is $\frac{1}{2}a\Delta t^2$, where a is the acceleration, so that, comparing the two expressions, we conclude that $\dfrac{v^2}{R}$ represents the acceleration a with which the Moon continuously falls toward the Earth, missing it all the time.

Thus we can write for that acceleration:

$$a = \frac{v^2}{R} = \left(\frac{v}{R}\right)^2 R = \omega^2 R$$

where

$$\omega = \frac{v}{R}$$

is the *angular velocity* of the Moon in its orbit. Angular velocity ω (the Greek letter *omega*) of any rotational motion is very simply connected with the rotation period T. In fact, we can rewrite the formula as:

$$\omega = \frac{2\pi v}{2\pi R} = 2\pi \frac{v}{s}$$

where $s = 2\pi R$ is the total length of the orbit. Apparently the rotation period T is equal to $\dfrac{s}{v}$ so that the formula becomes:

$$\omega = \frac{2\pi}{T}$$

The Moon takes 27.3 days, or $2.35 \cdot 10^6$ seconds, for a complete revolution around the Earth. Substituting this value for T in the expression, we get:

$$\omega = 2.67 \cdot 10^{-6} \frac{1}{\text{sec}}$$

Using this value for ω and taking $R = 384,400$ km $= 3.844 \cdot 10^{10}$ cm, Newton obtained for the acceleration of the falling Moon the value 0.27 cm/sec² which is *3640* times smaller than the acceleration 981 cm/sec² on the surface of the Earth. Thus, it became clear that the force of gravity decreases with the distance from the Earth, but what is the law governing this decrease? The falling apple is at the distance 6371 km from the center of the Earth and the Moon is at the distance 384,400 km, i.e., 60.1 times farther. Comparing the two ratios *3640* and *60.1*, Newton noticed that the first is almost exactly equal to the square of the second. This meant that the law of gravity is very simple: *the force of attraction decreases as the inverse square of the distance.*

But, if the Earth attracts the apple and the Moon, why not assume that the Sun attracts the Earth and other planets, keeping them in their orbits? And then there should also be an attraction between individual planets disturbing, in turn, their motions around the central body of the system. And, if so, two apples should also attract each other, though the force between them may be too weak to be noticed by our senses. Clearly this force of universal gravitational attraction must depend on the masses of the interacting bodies. According to one of the basic laws of mechanics formulated by Newton, *a given force acting on a certain material body communicates to this body an acceleration which is proportional to the force and inversely proportional to the mass of the body;* indeed, it takes twice as much effort to bring up to the same speed a body with double mass. Thus, from Galileo's finding that all bodies, independent of their weight, fall with the same acceleration in the field of gravity, one must conclude that the forces pulling them down are proportional to their mass; i.e., to the resistance to acceleration. And, if so, gravitational force might also be expected to be proportional to the mass of another body. Gravitational attraction between the Earth and the Moon is very large because both bodies are very massive. The attraction between the Earth and an apple is much weaker because the apple is so small, and the attraction between two apples must be quite negligible. By using arguments of that kind, Newton came to the formulation of the *Law of Universal Gravity,* according to which *every two material objects attract each other with a force proportional to the product of their masses, and inversely proportional to the square of the distance between them.* If we write M_1 and M_2 for the masses of two interacting bodies, and R for the distance between them, the force of gravitational interaction will be expressed by a simple formula:

$$F = \frac{GM_1M_2}{R^2}$$

where G (for gravity) is a universal constant.

Newton did not live to witness a direct experimental proof of his law of attraction between two bodies each not much larger than an apple, but three-quarters of a century after his death another talented Britisher, Henry Cavendish, demonstrated the proof beyond argument. In order to prove the existence of gravitational attraction between everyday-sized bodies, Cavendish used very delicate equipment that in his day represented the height of experimental skill but which can be found today in most physics lecture rooms to impress Newton's law of gravity on the minds of freshmen. The principle of the Cavendish balance is shown in Fig. 8. A light bar with two small spheres attached at each end is suspended on a long thread as thin as a cobweb, and placed inside a glass box to keep air currents from disturbing it. Outside the glass box are suspended two very massive spheres which can be rotated around the central axis. After the system comes to the state of equilibrium, the position of the large sphere is changed, and it is observed that the bar with the small spheres turns through a certain angle as a result of gravitational

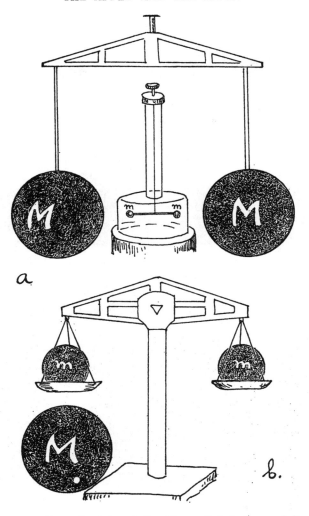

Fig. 8. The principle of the Cavendish balance (a), and (b) Boys' modification.

attraction to the large sphere. Measuring the deflection angle and knowing the resistance of the thread to a twist, Cavendish could estimate the force with which massive spheres acted on the little ones. From these experiments he found that the numerical value of the coefficient G in Newton's formula is 6.66×10^{-8} if the lengths, masses, and time are measured in centimeters, grams, and seconds. Using this value, one can calculate that the gravity force between two apples placed close to each other is equivalent to the weight of one-billionth of an ounce!

A modified form of the Cavendish experiment was performed later by the British physicist C. V. Boys (1855–1944).* After having balanced two equal weights on the scales (Fig. 8) he placed a massive sphere under one of the plates, and observed a slight deflection; the attraction of the terrestrial globe on that weight was augmented by the attraction of the massive sphere. The observed deflection permitted Boys to calculate the ratio of the mass of the sphere to the mass of the Earth; the Earth, he found, weighs 6.10^{24} kilograms (kg).

* Author of *Soap Bubbles and the Forces Which Mould Them.*

Chapter 3

CALCULUS

It may seem hard to understand that Newton, having obtained the basic ideas of Universal Gravity in the very beginning of his scientific career, should have withheld publication for about twenty years until he was able to present a complete mathematical formulation of the Theory of Universal Gravity in his famous book, *Philosophiae Naturalis Principia Mathematica*, published in 1687.

The reason for such a long delay was that, although Newton had clear ideas concerning the physical laws of gravity, he lacked the mathematical methods necessary for the development of all the consequences of his fundamental law of interaction between the material bodies. The mathematical knowledge of his time was inadequate for the solution of the problems which arose in connection with gravitational interaction between material bodies. For example, in the treatment of the Earth-Moon problem described in the previous chapter, Newton had to assume that the force of gravity is inversely proportional to the square of the distance between the *centers* of these two bodies. But when an apple is attracted by the terrestrial globe, the force pulling it down is composed of an infinite number of different forces caused by the attraction of rocks at various depths under the roots of the apple tree, by the rocks of the Himalayas and the Rocky Mountains, by the waters of the Pacific Ocean, and by the molten central iron core of the Earth. In order to make the previously given derivation of the ratio of forces with which the Earth acts on the apple and on the Moon mathematically immaculate, Newton had to prove that all these forces add up to a single force which would be present if all the mass of the Earth were concentrated in its center.

This problem, similar to but much more complicated than Galileo's problem concerning the motion of a particle with constantly increasing velocity, was beyond the mathematical resources of Newton's time, and he had to develop his own mathematics. In doing so he laid the foundation for what is now known as *the calculus of infinitesimals,* or, simply, *Calculus.* This branch of mathematics, which is today an absolute "must" in the study of all physical sciences and is becoming more and more important in biology and other fields, differs from classical mathematical disciplines by using a method in which the lines, the surfaces, and the volumes of classical geometry are divided into a very large number of very small parts, and one considers the interrelations in the limiting case when the size of each subdivision goes to zero. We already have encountered such kinds of argumentation in Newton's derivation of the acceleration of the Moon (p. 39) where the second term on the left side of the equation can be neglected as compared with the first term if we consider the change of the Moon's position during a vanishingly short time interval. Let us consider a general kind of motion in which the coordinate x of a moving object is given as a function of time, t. In everyday language this means that the value of x changes in some regular way as the value of t changes. In the simplest case x may be proportional to t, and we write:

$$x = At$$

in which the A is a constant that makes the two sides of the equation equal.

This case is trivial. We take two moments of time t and $t + \Delta t$ where Δt is a small increment which is later to be made equal to zero. The distance traveled during this time interval is apparently:

$$A(t + \Delta t) - At = A\Delta t$$

and, dividing it by Δt we get exactly A. In this case we do not even need to make Δt infinitesimally small since it cancels out of the equation. Thus, we get for the time rate of change of x, or the "fluxion of x," as Newton called it:

$$\dot{x} = A$$

where a dot placed above the variable denotes its rate of change.

Let us now take a somewhat more complicated case given by:

$$x = At^2$$

Taking again the values of x for t and $t + \Delta t$, we obtain:

$$A(t + \Delta t)^2 - At^2$$

and, opening the parenthesis gives:

$$At^2 + 2At\Delta t + \Delta t^2 - At^2 = 2At\Delta t + \Delta t^2$$

Dividing this by Δt, we get a two-term expression:

$$2At + \Delta t$$

When Δt becomes infinitesimally small the last term

disappears and we have for the fluxion of $x = At^2$:

$$\dot{x} = 2At$$

Turning to the case of

$$x = At^3$$

we have to calculate the expression:

$$A(t + \Delta t)^3 - At^3$$

Multiplying $(t + \Delta t)$ by itself three times and subtracting At^3, we get:

$$A(t^3 + 3t^2\Delta t + 3t\Delta t^2 + \Delta t^3) - At^3 =$$
$$3At^2\Delta t + 3At\Delta t^2 + A\Delta t^3$$

and dividing by Δt:

$$3At^2 + 3At\Delta t + A\Delta t^2$$

When Δt becomes infinitesimally small, the last two terms vanish and we obtain for the fluxion of $x = At^3$:

$$\dot{x} = 3At^2$$

We can go on with $x = At^4$, $x = At^5$, etc., obtaining the fluxions $4At^3$, $5At^4$, etc. It is easy to notice the general rule: *the fluxion of $x = At^n$ where n is an integer number is equal to nAt^{n-1}.*

In the foregoing examples we calculate the fluxions of the quantities which are changing in direct proportion to time, to the square of time, the cube of time, etc. But what about quantities which change in *inverse* proportion to various powers of time? We know from algebra that:

$$t^{-1} = \frac{1}{t}; \qquad t^{-2} = \frac{1}{t^2}; \qquad t^{-3} = \frac{1}{t^3}; \quad \text{etc.}$$

Using these negative exponents and proceeding as

before, we find that the fluxions of

$$x = At^{-1}; \qquad x = At^{-2}; \qquad x = At^{-3}; \quad \text{etc.}$$

are:

$$\dot{x} = -At^{-2}; \qquad \dot{x} = -2At^{-3}; \qquad \dot{x} = -3At^{-4}; \quad \text{etc.}$$

The minus sign here stands because, in the case of *inverse* proportionality, variable quantities *decrease* with time, and the rate of change is *negative*. But the general rule for calculating the fluxions remains the same as in the case of direct proportionality: to get the expression of the fluxion we *multiply the original power function by its exponent, and reduce the value of the exponent by one unit*. The results of the foregoing discussion are summarized in a table below:

$x =$	At^{-3};	At^{-2};	At^{-1}; At; At^2; At^3; At^4; etc.
$\dot{x} =$	$-3At^{-4}$;	$-2At^{-3}$;	$-At^{-2}$; A; $2At$; $3At^2$; $4At^3$; etc.

While $\dot{x}$ in Newton's notations represents *the rate of change* of x, $\ddot{x}$ represents *the rate of change of this rate of change*. Thus, for example, if $x = At^3$,

$$\dot{x} = 3At^2 \qquad \text{and} \qquad \ddot{x} = \boxed{\dot{3At^2}} = 3A \cdot 2t = 6At$$

Similarly $\dddot{x}$, which is *the rate of change of the rate of change of the rate of change*, will be, in the same case:

$$\dddot{x} = \boxed{\dot{6At}} = 6A$$

We can now try these simple rules on Galileo's formula for the free fall of material bodies. In Chapter 1 we found that the distance s covered at the time t is given by

$$s = \frac{1}{2} at^2$$

Since the velocity v is the rate of change of position, we have:

$$v = \dot{s} = \frac{1}{2} a \cdot 2t = at$$

which says that the velocity is simply proportional to time. For the acceleration a, which is the rate of change of velocity (or the rate of change of the rate of change of position), we have:

$$a = \ddot{s} = \dot{v} = a$$

which is, of course, a trivial result.

Before we leave this subject, we must notice that Newton's fluxion notations are very seldom used in the books of today. At the same time that Newton was developing his method of fluxions now known as differential calculus, a German mathematician, Gottfried W. Leibniz, was working along the same lines using, however, a somewhat different terminology and system of notations. What Newton called first, second, etc., fluxions, Leibniz called first, second, etc., *derivatives*, and, instead of writing $\dot{x}$; $\ddot{x}$; $\dddot{x}$; etc., he wrote:

$$\frac{dx}{dt}; \quad \frac{d^2x}{dt^2}; \quad \frac{d^3x}{dt^3}; \quad \text{etc.}$$

But the mathematical content of the two systems is, of course, the same.

While differential calculus considers the relation between the parts of geometrical figures when these parts become infinitely small, *integral calculus* has an exactly opposite task: the integration of infinitely small parts into geometrical figures of final size. We encountered this method in Chapter 1 when we described Galileo's method of adding up a very large number of very thin rectangles, the area of which represented the motion of a particle during a very short time interval. Similar methods were used before Galileo by Greek mathematicians for finding volumes of cones and other

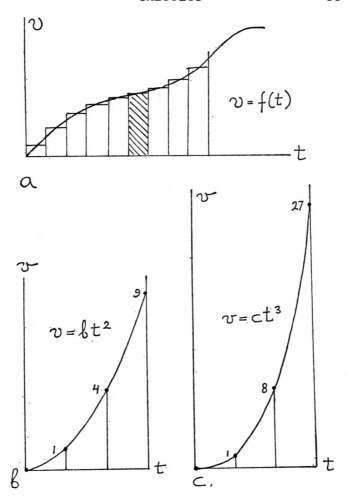

Fig. 9. Integration of (a) arbitrary function; (b) quadratic function; (c) cubic function.

simple geometrical figures, but the general method for solving such kinds of problems was not known.

To understand the relation between differential and integral calculus, let us consider the motion of a point whose velocity is given by the function $v(t)$ as shown in Fig. 9. Using the same arguments as in the simple case represented in Fig. 3, we conclude that the distance s traveled during the time t is given by the *area* under the velocity curve. The rate of change of s at any particular moment is given by the velocity of motion at that moment, so that we can write:

$$\dot{s} = v \qquad \text{or} \qquad \frac{ds}{dt} = v$$

in Newton's and Leibniz' notations, respectively. Thus, if v is given as the function of time, s must be such a function of time that its fluxion (or derivative) is equal to v. In the case of uniformly accelerated motion:

$$v = at$$

so that we have to find a function of time the fluxion of which is equal to at. Consulting the table on p. 53, we find that the fluxion of At^2 is $2At$, so that the derivative of $\frac{1}{2}At^2$ is equal to At. Thus, writing a instead of A, we find that $s = \frac{1}{2}at^2$. This is, of course, the same result that Galileo had obtained from purely geometrical considerations.

But let us consider two more complicated cases, one in which the velocity increases as the square of time and another in which it increases as the cube of time. For these two cases we must write:

$$v = bt^2 \qquad \text{and} \qquad v = ct^3$$

These two cases are represented by graphs in Fig. 9, and, just as in the previous simple example, the distances traveled are represented by the areas under the curves. But, since we have here curving and not straight lines, there is no simple geometrical rule showing how to find these areas. Using Newton's method, we look again into the table on p. 53 and find that the derivatives of At^3 and At^4 are $3At^2$ and $4At^3$, differing from the given expressions of the velocities only by the numerical coefficients. Thus, putting $3A = b$ and $4A = c$, we find for the areas under the two curves:

$$s_b = \frac{1}{3} bt^3 \quad \text{and} \quad s_c = \frac{1}{4} ct^4$$

The method is quite general, and can be used for any power of t, as well as for more complicated expressions such as:

$$v = at + bt^2 + ct^3$$

for which we get:

$$s = \frac{1}{2} at^2 + \frac{1}{3} bt^3 + \frac{1}{4} ct^4$$

From the discussion we see that integral calculus is a *reverse* of differential calculus: *the problem here is to find the unknown function whose derivative is equal to a given function.* Thus, we can now rewrite the table on p. 53, changing the order of the two lines, and changing the numerical coefficients, in the form:

$\dot{x} =$	At^{-4}; At^{-3}; At^{-2}; A; At; At^2; At^3; etc.
$x =$	$-\dfrac{A}{3} t^{-3}; -\dfrac{A}{2} t^{-2}; -At^{-1}; At; \dfrac{A}{2} t^2; \dfrac{A}{3} t^3; \dfrac{A}{4} t^4;$ etc.

We say that x is an integral of $\dot{x}$. In Newton's notations one writes:

$$x = (\dot{x})'$$

where the prime accent placed outside the parenthesis counteracts the *dot* above the x. In Leibniz' notations, we write:

$$x = \int \dot{x} dt$$

where the symbol in the front on the right side is nothing but an elongated S standing for the word *sum*.

Let us apply this new table to the same old example of a uniformly accelerated motion. Since acceleration is constant, we write:

$$\ddot{x} = a \qquad \text{or} \qquad \boxed{\dot{\dot{x}}} = a$$

from which it follows that:

$$\dot{x} = \int a\,dt = at$$

Integrating a second time and consulting our new table, we obtain:

$$x = \int at \cdot dt = \frac{a}{2}t^2$$

i.e., the same result as obtained before. If acceleration is not constant but, let us say, proportional to time, we have:

$$\ddot{x} = bt$$

$$\dot{x} = \int bt\,dt = \frac{1}{2}bt^2$$

$$x = \int \frac{1}{2}bt^2 dt = \frac{b}{2}\int t^2 dt = \frac{b}{6}t^3$$

Thus, in this case the distance covered by a moving object would increase as the cube of time.

The elementary formulation of differential and integral calculus can be extended into three dimensions, when all three coordinates x, y, and z are present, but this we leave to the reader who found the previous discussion too easy.

Having developed the basic principles of calculus, Newton applied them to the solution of problems which stood in the way of his Theory of Universal Gravity— in the first place, to the problem of the gravity force

exerted by the body of the Earth on any small material
object at any distance from its center. For this purpose
he divided the Earth into thin concentric shells, and

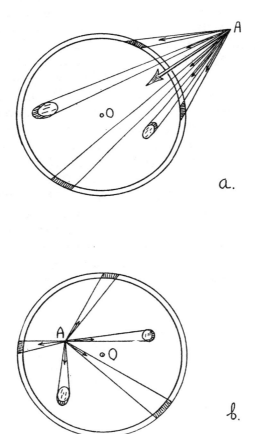

Fig. 10. (a) Gravitational force exerted by spherical
shell on the outside point; (b) same thing, only inside
instead of outside.

considered their gravitational action separately (Fig.
10). In order to use integral calculus we must divide
the surfaces of the shells into a large number of small

regions of equal areas, and then calculate the gravitational force exerted by each region on the object O in accordance with the inverse square law. This analysis leads to a very large number of force vectors applied to the point O, and these vectors should be integrated according to the rules of vector addition. The actual solution of that problem is beyond the elementary principles we have discussed, but Newton managed to solve it. The result was that when the point O was *outside* the spherical shell, all the vectors added up to form a single vector *equal to the gravity force which would exist if the entire mass of the spherical shell were concentrated in its center.* In the case where point O was *inside* the shell, the sum of all vectors was exactly zero, so that *no gravity force was acting on the object.* This result solved Newton's trouble concerning the attractive force exerted by the Earth on an apple, and justified the Law of Universal Gravity he had stated when he was a young man contemplating the riddles of nature in the orchard on the Lincolnshire farm.

Chapter 4

PLANETARY ORBITS

Now that we have learned a little bit of calculus, we can try to apply it to the motion of natural and artificial celestial bodies under the force of gravity. Let us first calculate how fast a rocket should be shot from the surface of the Earth in order to escape the bond of terrestrial gravity. Consider furniture movers who have to move a grand piano from the street level to a certain floor in a high apartment building. Everybody will agree (and especially the furniture movers) that to bring a grand piano up three floors takes three times more work than to bring it up one floor. The work of carrying heavy pieces of furniture is also proportional to their weight, and in carrying up six chairs one does six times more work than in carrying up just one chair.

This is all very inconsequential, of course, but what about the work necessary to raise a rocket sufficiently high to put it into a prescribed orbit, or the work of bringing it still higher so that it would drop down on the Moon? In solving problems of this kind, we must remember that the force of gravity decreases with the distance from the center of the Earth; the higher we lift the object the easier it will be to lift it still higher.

Fig. 11 shows the change of gravitational force with the distance from the center of the Earth. In order to calculate the total work needed to bring an object from the surface of the Earth (distance R_o from the center) to the distance R, taking into account the continuous decrease of gravitational force, we divide the

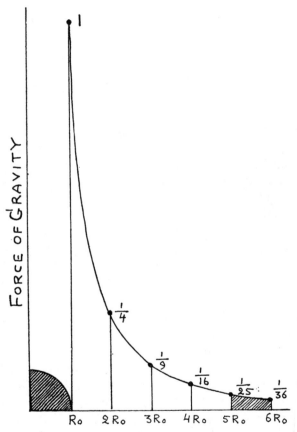

Fig. 11. Decrease of gravitational force with distance (R_0 is the radius of the Earth).

distance from R_0 to R into a large number of small intervals *dr,* and consider the work done while covering that distance. Since for a small variation of distance the force of gravity can be considered as practically constant (remember the furniture movers), the work done is the product of the force moving the object by the distance moved; i.e., the area of the dashed rectangle in Fig. 11. Going to the limit of infinitely small displacements, we conclude that the total work of lifting an object from R_0 to R is the area under the curve representing the force of attraction, or, in the notations of the previous chapter, the integral:

$$W = \int_{R_0}^{R} \frac{GMm}{r^2}\, dr = GMm \int_{R_0}^{R} \frac{1}{r^2}\, dr$$

(Since constants are not affected in the process of integration, we can remove GMm from under the integral sign and multiply it with the final result of integration.) Looking into the table on p. 57, we find that the integral of $\frac{1}{r^2}$ is $-\frac{1}{r}$ $\left(\text{since the derivative of } \frac{1}{r} \text{ is } -\frac{1}{r^2}\right)$. Thus the work done is:

$$W = -\frac{GMm}{R} - \left(-\frac{GMm}{R_0}\right) = GMm\left(\frac{1}{R_0} - \frac{1}{R}\right)$$

The expression

$$P_R = -\frac{GM}{R}$$

(referring to the unit mass to be lifted) is known as *gravitational potential*, and we say that the work done to lift a unit mass from the surface of the Earth to a certain distance out in space is equal to the difference of the gravitational potential at these two places.

Such simple considerations were known to Newton in the very early stages of his studies, but he faced a much more difficult job of explaining the exact laws of the motion of planets around the Sun, and of the planetary satellites—laws that were discovered more than half a century before Newton by a German astronomer,

Johannes Kepler. In studying the motion of planets in respect to fixed stars, Kepler had data obtained by his teacher, Tycho Brahe. Kepler found that *the orbits of all planets are ellipses with the Sun located in one of the two foci.* The ancient Greek mathematicians defined the ellipse as the cross section of a cone cut by a plane inclined to the cone's axis; the larger the inclination of the plane the more elongated the ellipse. If the plane is perpendicular to the axis, the ellipse degenerates into a circle. Another equivalent definition of an ellipse is a closed curve having the property that the sum of the distances of each of its points from two fixed points on the longer axis, the *foci,* is always the same. This definition gives a convenient method of drawing an ellipse by means of two pins and a string as shown in Fig. 12.

The second law of Kepler states that *the motion of planets along their elliptical orbits proceeds in such a way that an imaginary line connecting the Sun with the planet sweeps over equal areas of the planetary orbit in equal intervals of time* (Fig. 12).

Finally, the third law, which Kepler published nine years later, states that *the square of periods of revolution of different planets stand in the same ratio as the cubes of their mean distances from the Sun.* Thus, for example, the distances of Mercury, Venus, Mars, and Jupiter, expressed in terms of the distance of the Earth from the Sun (the so-called "astronomical unit" of distance), are 0.387; 0.723; 1.524; 5.203, while their rotation periods are 0.241; 0.615; 1.881, and 11.860 years, respectively. Taking the cubes of the first sequence of numbers (distances) and the squares of the second sequence (periods), we obtain identical numerical results, namely: 0.0580; 0.3785; 3.5396; and 140.85.

In his early studies Newton considered, for simplicity, the orbit of the Moon to be exactly circular, and this approximation led him to the comparatively elementary derivation of the law of gravity as presented in Chapter 2. But, having made this first step, he had to prove that if the Law of Universal Gravity is exactly correct, plane-

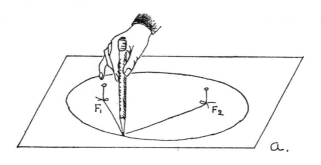

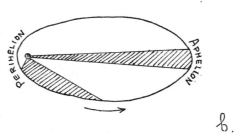

Fig. 12. (a) A simple way to draw an ellipse; (b) the second law of Kepler.

tary orbits deviating from circles must be ellipses with the Sun in one of the foci. The same goes, of course, for the Moon, since its orbit is not exactly a circle but an ellipse. Newton could not establish a proof by the classical geometry of circles and straight lines exclusively, and, as discussed earlier, he developed the differential calculus essentially for the purpose of dealing with that problem. The elements of differential calculus given in the previous chapter do not suffice to reproduce Newton's proof that planetary orbits should be ellipses, but it is hoped that this discussion will help the reader at least to understand *how* Newton solved that problem. In Fig. 13 we show the motion of a planet

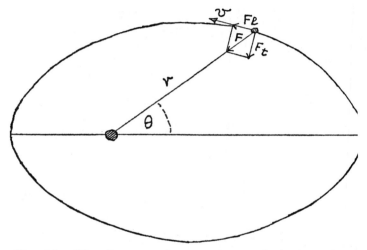

Fig. 13. The forces acting on the motion of a planet along its elliptical trajectory.

along a certain trajectory OO' with a certain velocity v. For motions of such kinds it is convenient to describe the position of the planet at any moment by giving its distance r from the Sun, and the angle θ (theta) which the line drawn from the Sun to the planet (radius vector) forms with some fixed direction in space, say the direction to some fixed star somewhere in the plane of the ecliptic. While the position of the planet is given by the coordinates r and theta, the rate of change of its position is given by the fluxions $\dot{r}$ and $\dot{\theta}$, and the rate of change of the rate of change (i.e. acceleration) by the

second fluxions: $\ddot{r}$ and $\ddot{\theta}$. The gravity force $F = \dfrac{GMm}{r^2}$
acting on the planet is, generally speaking, not perpendicular to its orbit as it would be in circular motion. Thus, using the rule of addition of forces, we can break the motion up into two components: one, F_l directed along the orbit, and another, F_t perpendicular to it.*

Having done this, and using Newton's basic law of mechanics, which states that acceleration of motion in any direction is proportional to the component of the force acting in that direction, one obtains the so-called *differential equations* of the motion of the planet. These equations give the relations between coordinates r and θ, their fluxions $\dot{r}$ and $\dot{\theta}$, and their second fluxions $\ddot{r}$ and $\ddot{\theta}$. The rest is nothing but pure mathematics— just finding how r and θ must depend on time in order that their first and second fluxions, as well as themselves, satisfy the differential equations. And the answer is that motion must proceed along an ellipse with the Sun in a focus in such a way that the radius vector sweeps equal areas during equal intervals of time.

While we were able here to give only a "descriptive" derivation of the first two laws of Kepler, we can give an exact derivation of his third law, making a simplifying assumption that planetary orbits are circular. In fact, we have seen in Chapter 2 that the centripetal (directed toward a center) acceleration of circular motion is v^2/R where v is the velocity of the moving body and R the radius of the orbit. Since the centripetal acceleration multiplied by mass must be equal to the force of gravitational attraction, we may write:

$$\frac{mv^2}{R} = \frac{GMm}{R^2}$$

* Indices l and t stand for the words *longitudinal* and *transversal*.

On the other hand, since the length of the circular orbit is $2\pi R$, the period T of one revolution is apparently given by the formula:

$$T = \frac{2\pi R}{v}$$

from which follows

$$v = \frac{2\pi R}{T}$$

Substituting this value of v in the first equation we obtain:

$$\frac{m4\pi^2 R^2}{T^2 R} = \frac{GMm}{R^2}$$

or, rearranging and canceling m on both sides:

$$4\pi^2 R^3 = GMT^2$$

Well, that's all there is to it! The formula says that *the cubes of R* are *proportional to the squares of T*, which is exactly the third law of Kepler.

By a more elaborate application of calculus one can show that the same law holds also for a more general case of elliptical orbits.

Thus, inventing mathematics necessary for the solution of his problem, Newton was able to show that the motion of the members of the solar family does obey his Law of Universal Gravity.

Chapter 5

THE EARTH AS A SPINNING TOP

Having solved the problem of how the forces of terrestrial gravity hold the Moon in its orbit, and how the gravity of the Sun makes the Earth, as well as other planets, move around it along an elliptical trajectory, Newton turned his attention to the question of the influence which these two celestial bodies exert on the rotation of our globe about its axis. He realized that because of axial rotation the Earth must have the shape of a compressed spheroid, since the gravity at equatorial regions is partially compensated by centrifugal force. Indeed, the equatorial radius of the Earth is 13 miles longer than its polar radius, and acceleration of gravity at the equator is 0.3 per cent less than at the poles. Thus, the Earth can be considered as a sphere surrounded by an equatorial bulge (shaded area in the lower part of Fig. 14), which is about 13 miles thick at the equator, coming down to zero at the poles. While the gravitational forces of the Sun and the Moon acting on the material of the spherical part of the Earth are equivalent to a single force applied in the center, the forces acting on the equatorial bulge do not balance that way. Indeed, since gravity decreases with distance, the force F_1 acting on the part of the bulge which is turned to the attracting body (Sun or Moon) is larger than the force F_2 acting on the opposite side. As a result, there appears a *torque* or twist-force, tending to straighten the rotational axis of the Earth, making it perpendicular to the plane of the Earth's orbit (ecliptic)

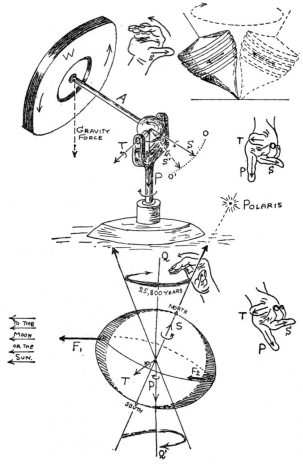

Fig. 14.   Spinning gyro and spinning Earth.

or the plane of the orbit of the Moon. Why then does the rotational axis of the Earth not turn in the way it should under the action of these forces?

To answer this question, we have to realize that our globe is actually a giant spinning top, in its movement like that amusing toy familiar to us all since childhood. When set up in a fast rotation, the top does not fall down, as it seemingly should, but maintains an inclined position in respect to the floor, and while it spins, the rotation axis describes a wide cone around the vertical (upper right corner in Fig. 14). Only when, because of friction, the spinning of the top slows down, will it fall on the floor and roll under the sofa. A little more elaborate model of the spinning top, used in classes on theoretical mechanics, is shown in the upper half of Fig. 14. It consists of a fork F which can rotate around a vertical axis, supporting a bar A which can move freely up and down around the suspension point. At the free end of the bar is attached a flywheel W which rotates with slight friction on ball bearings. If the wheel is not in motion, the normal position of the system will be with the bar sloping down and the wheel resting on the table. If, however, we set the wheel in fast rotation, things will run in an entirely different way, a way almost unbelievable to a person who observes this phenomenon for the first time. The bar and the wheel will not fall down and, as long as the wheel spins, the wheel, the bar, and the forked support will rotate slowly around the vertical axis. This is the well-known principle of the *gyroscope,* which has many practical applications, among them the "gyroscopic compass," steering ships across the ocean and planes through air, and the "gyroscopic stabilizers," preventing roll and yaw in rough weather.

Probably the most amusing application of the gyroscope was made by a French physicist, Jean Perrin, who packed a running aviation-gyro into a suitcase and checked it at the Paris railroad station (commercial airlines did not exist at that time). When the French equivalent of a redcap picked up the suitcase and, walking through the station, tried to turn a corner, the suitcase he carried refused to go along. When the astonished redcap applied force, the suitcase turned

Fig. 15. Perrin's experiment.

on its handle at an unexpected angle, twisting the redcap's wrist (Fig. 15). Shouting in French, "The Devil himself must be inside!" the redcap dropped the suitcase and ran away. A year later Jean Perrin received the Nobel Prize, not for his gyro-experimentation but for his work on the thermal motion of molecules.

To understand the peculiar behavior of gyroscopes, we must familiarize ourselves with the vector representation of rotational motion. In Chapter 1 we saw that the velocity of translatory motion can be represented by an arrow (vector) drawn in the direction of the motion and having a length proportional to the velocity. For rotation a similar method is used. One draws the arrow along the rotation axis, the length of the arrow corresponding to the angular velocity measured in RPM (revolutions per minute) or any other equivalent unit. The way in which the arrow points is settled by the "right-hand screw" convention: *if you put the bent fingers of the right hand in the direction of rotation, your thumb will show the correct direction of the arrow.* (This rule is also quite handy when you try to unscrew the top of a glass jar or anything else.) In the upper part of Fig. 14, vector *S* shows the rotational velocity of the flywheel. The torque (the twist force) due to gravity is shown by vector *T* running across the fork's hinges. Extending the laws of translatory motion for the case of rotational motion, we would expect that the rate of change of the velocity is proportional to the applied torque. Thus, the effect of gravity on a spinning top will be the change of rotational velocity given by vector *S*, to that given by vector *S′*, i.e., the rotation around the vertical axis. And that is exactly what is observed in the behavior of a spinning top.

The space relation between the angular velocity of the flywheel, the torque, and the resulting motion is shown by hands in Fig. 14. If you point the middle finger of your right hand in the direction of the rotation vector and the thumb in the direction of the torque vector, the index finger will indicate the resulting rotation of the system.

The phenomenon we have just described is known as *precession* and is common to all rotating bodies, be they stars or planets, children's toys, or electrons in an atom. In the motion of the Earth precession is caused by gravitational attraction of the Sun and the Moon, the latter playing a primary role since, being less massive than the Sun, it is much closer to the Earth. The combined effect of lunar-solar precession turns the axis of the Earth by 50 angular seconds per year, and causes it to complete a full circle every 25,800 years. This phenomenon, which is responsible for slow changes of the dates on which the spring and fall begin (precession of equinoxes), was discovered by the Greek astronomer Hipparchus about 125 B.C., but the explanation had to wait until Newton's formulation of the Theory of Universal Gravity.

Chapter 6

THE TIDES

Another and a more important influence of the Sun and the Moon on the Earth is diurnal deformation of the Earth's body, most noticeable in the phenomenon of ocean tides. Newton realized that periodic rise and fall of the ocean level results from gravitational attraction exerted by the Sun and the Moon on the ocean waters, the influence of the Moon being considerably greater because, even though it is much smaller than the Sun, it is much, much closer to us. He argued that since gravity forces decrease with distance, the pull exerted on ocean water on the moonlit or sunlit side of our globe is greater than that on the opposite side, and thus must lift the water above the normal level.

Many people who hear for the first time this explanation of ocean tides find it hard to understand why there are two tidal waves, one on the side turned toward the Moon or the Sun, and another on the opposite side where ocean waters seem to move in the direction opposite to the gravitational pull. To explain this we must discuss in some detail the dynamics of the Sun-Earth-Moon system. If the Moon were fixed in a given position, sitting on the top of a giant tower erected on some part of the Earth's surface, or if the Earth itself were kept stationary at some point of its orbit by some supernatural force, the ocean waters would indeed collect on one side and there would be a lowering of ocean level on the opposite side. But since the Moon revolves around the Earth, and the Earth revolves around the Sun, the situation is quite different.

Let us consider first the solar tides. Since in its motion around the Sun the Earth stays in one piece, the linear velocity of the side turned toward the Sun (F in Fig. 16a) is less than the linear velocity of the center (C) of the Earth, which is in its turn lower than the linear velocity of the rear side (R). On the other hand, as we have seen in Chapter 4, the linear velocities of circular orbital motion under the action of solar gravity must decrease with the distance from the Sun. Thus, point F has less linear velocity than is needed to maintain circular motion, and so will have the tendency to be deflected toward the Sun, as indicated by a dotted arrow at F in Fig. 16a. Similarly, the point R has higher linear velocity than needed for its circular orbit, and will have a tendency to move farther away from the Sun (dotted arrow at R). Thus, if there were no attraction between different parts of the material forming the Earth, it would be broken to pieces, which would be spread in the form of a broad disc all over the plane of the ecliptic. This does not happen, however, because the gravitational attraction G between different parts of the Earth tends to hold it together. As a compromise, our globe becomes elongated in the direction of the orbital radius with two bulges on each side.

Regarding lunar tides the argument is exactly the same if one remembers that both the Earth and the Moon move around the common center of gravity. Since the Moon is about eighty times lighter than the Earth, the common center of gravity between the two bodies is one-eightieth of the distance from the center of the Earth. Remembering that this distance equals sixty Earth radii, we conclude that the center of gravity

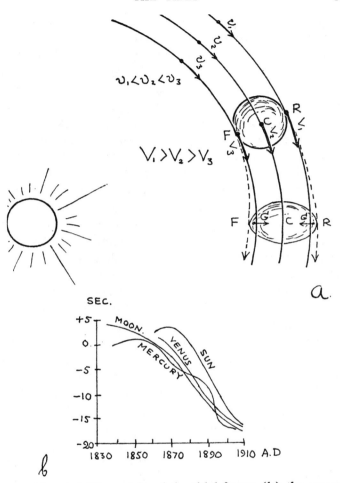

$\upsilon_1 < \upsilon_2 < \upsilon_3$

$V_1 > V_2 > V_3$

SEC.

Fig. 16. (a) The origin of the tidal force; (b) the apparent delay in the motion of celestial bodies.

of the Earth-Moon system is located at $60/80 = 3/4$ of the Earth's radius from its center. In spite of a quantitative difference in geometry, the physical argument remains the same. The waters of the Earth's oceans form two bulges, one directed to the common center of gravity (which is also the direction toward the Moon) and the other in the opposite direction.

When Sun, Earth, and Moon are located on one straight line—i.e., during the new- and full-moon periods—the tidal action of the Moon and the Sun add up and the tides are especially high. During the first and last quarters, however, lunar high tides coincide with solar low tides, and the total effect is reduced.

Since the Earth is not absolutely rigid, the lunar-solar tidal forces deform its body, though these deformations are considerably smaller than those in the liquid envelope. The American physicist A. A. Michelson* found from his experiments that every twelve hours the surface of the Earth is deformed about one foot, as compared with a four- to five-foot deformation of the ocean's surface. Since the deformations of the Earth's crust occur slowly and smoothly, we do not realize that we live on a rocking foundation, but when we observe the ocean tide rising at the shores of the continents, we must remember that what we see is only the *difference* between the vertical motion of land and water.

Ocean tides running around our globe experience friction at the ocean bottom (especially in shallow basins such as the Bering Sea) and also lose energy colliding with continental shore lines. Two British scientists, Sir Harold Jeffreys and Sir Geoffrey Taylor, estimated that the total work done continuously by the tides amounts to about two billion horsepower. As a result of this dissipation of energy, the Earth slows in its rotation about its axis, just as an automobile's wheels do when brakes are applied. Comparing this loss of energy in tides with the total energy of the Earth's rotation, one finds that it slows down by 0.00000002 second per rotation; each day is two hundred millionths

* *Michelson and the Speed of Light,* Bernard Jaffe, Science Study Series, 1960.

of a second longer than the previous one. This is a very small change, and there is no way of measuring it from today to tomorrow, or from New Year to New Year. But the effect accumulates as the years go by. One hundred years contain 36,525 days, so that a century ago days were 0.0007 second shorter than now. On the average, between then and now, the length of the day was 0.00035 second shorter than at present. But, since 36,525 days have passed by, the total accumulated error must be: $36,525 \times 0.00035 = 14$ seconds.

Fourteen seconds per century is a small figure, but it is well within the accuracy of astronomical observations and calculations. In fact, this slowing down of the rotation of the Earth about its axis explains a discrepancy which puzzled astronomers for a long time. Indeed, comparing the positions of the Sun, Moon, Mercury, and Venus in respect to the fixed stars, astronomers noticed that they seemed systematically *ahead of time* as compared with the position calculated a century ago on the basis of celestial mechanics (Fig. 16b). If a TV program starts fifteen minutes earlier than you expect it to start, if you find a store closed when you arrive less than fifteen minutes before closing time, and if you miss a train when you were sure you would catch it, you should not blame the radio station, the shop, and the railroad, but you should blame your watch. It is probably fifteen minutes slow. Similarly, the discrepancy of fifteen seconds in timing astronomical events should be ascribed to the slowing down of the Earth and not to the speeding up of all celestial bodies. Until the slowing down of the Earth's rotation was realized, astronomers used the Earth as the perfect clock. Now they know better and introduce the correction caused by tidal friction.

Early in this century the British astronomer George Darwin, son of the famous author of *Origin of the*

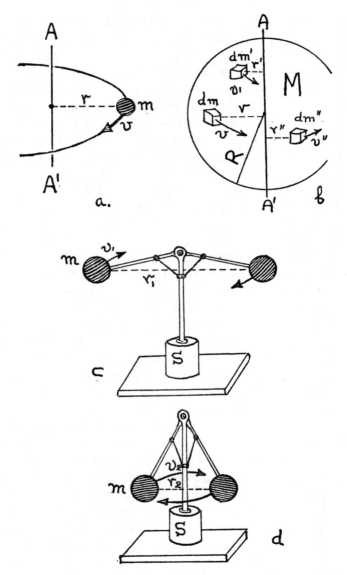

Fig. 17. The angular momentum of a rotating or revolving body is (a) the product of the body's mass (*m*), velocity (*v*), and its distance from the axis of rotation (*r*).

Species, undertook a study of the problem of how, in the long run, the loss of energy through tidal friction affects the Earth-Moon system.

In order to understand Darwin's argument we have to be acquainted with an important mechanical quantity known as the *angular momentum* of a revolving or a rotating material body. Let us consider a mass m revolving with the velocity v around a fixed axis AA' at a distance r from it (Fig. 17a). This may be the Earth revolving around the Sun, the Moon revolving around the Earth, or just a stone tied to a string in the hand of a boy swinging it around. Angular momentum I is defined as the product of the mass of the body, its velocity, and its distance from the axis:

$$I = mvr$$

The situation becomes a little more complicated when we consider a material body, be it a flywheel or the Earth, that rotates around an axis passing through the body's center (Fig. 17b). While in the previous case all parts of the body move with about the same velocity (as long as the size of the body is small as compared with the size of the orbit), various parts of a body rotating around an axis passing through its center have quite different velocities; the farther away a part of the body is from the rotation axis, the faster it moves. In the case of the Earth, for example, the points at the equator have much greater velocities than the points at the Arctic and Antarctic circles, and the points on the poles do not move at all. How then can we define

Calculation of the angular momentum of a rotating rigid body (b) is done by summing the angular momenta of an infinite number of small pieces, such as dm, dm', dm'', etc. Change in velocity to conserve angular momentum is illustrated in (c) and (d).

angular momentum in such a case? The way to do it, of course, is to use the integral calculus.

We divide the entire mass m of the body into a large number of small pieces dm, dm', dm'', etc., and calculate the angular momentum for each of them. Three such pieces shown in the figure are located at the distances r, r', and r'' from the axis and have the velocities v, v', and v'', which are, of course, proportional to these distances. In order to obtain the angular momentum I of the entire body, we have to *integrate* the angular momenta of all the pieces by writing:

$$I = \int dm_i v_i r_i$$

where the integration is extended over the entire body. Using the calculus, one can show that

$$I = \frac{2}{5} v_r r$$

where r is the radius of the rotating body and v_r is the velocity of the points at its equator.

One of the basic laws of the classical mechanics derived from Newton is the law of *the conservation of angular momentum*, which states that if we have any number of bodies rotating around their axes, and revolving around one another, *the total angular momentum of the system must always remain constant.*

An elementary classroom demonstration of that law can be carried out by using a gadget shown in the lower part of Fig. 17. It consists of two weights at the ends of two rods attached to the top of a vertical axis which can rotate with very little friction in a socket S. A special device (not shown in the picture) permits us at will to lift the balls up (Fig. 17c) or to bring them down (Fig. 17d).

Suppose, having the weights in the elevated position (c), we spin the system around its axis, thus communicating to it a certain amount of angular momentum. The angular momentum of each ball will be, according to the previous definition, equal to $m v_1 r_1$ where v_1 and r_1 have the meaning indicated in Fig. 17c. As the system is spinning, we lower the balls to the position indicated in Fig. 17d, so that their new distance r_2 from the axis becomes one-half of the previous distance r_1. Since $m v r$ must not change, the decrease of r by a factor of 2 must result in an increase of v by the same factor. Thus, the law of conservation of angular momentum requires that the velocities must be doubled and, indeed, one observes in the second case that $v_2 = 2 v_1$.

This principle is used for the purpose of producing astounding effects by circus acrobats, Ice Follies skaters, etc. Rotating on a rope or on the ice surface at comparatively low speed with the hands extended laterally in both directions, they suddenly bring their hands close to their bodies and become glittering whirlpools.

Returning to the Earth-Moon system, we conclude that the law of the conservation of angular momentum requires that the slowing down of the rotation of the Earth around its axis caused by tidal friction must result in an equal increase of angular momentum of the Moon in its orbital motion around the Earth.

How will this increase of angular momentum affect the motion of the Moon? The angular momentum of the Moon's orbital motion is:

$$I = m v r$$

where m is the mass of the Moon, v its velocity, and r the radius of the orbit. On the other hand, Newton's law of gravity, combined with the formula of centrifugal force, gives us:

$$\frac{GMm}{r^2} = \frac{mv^2}{r}$$

where M is the mass of the Earth. Thus:

$$\frac{GM}{r} = v^2$$

From this, and the above expression for I, follows:

$$r = \frac{I^2}{GMm^2}$$

and:

$$v = \frac{GMm}{I}$$

and the reader can take the author's word for it, if he is unable to reproduce the derivation. It follows from the above formulas: *the increase of the angular momentum of the Moon in its motion around the Earth must result in the increase of its distance from the Earth and the decrease of its linear velocity.*

From the observed slowing down of the Earth's rotation one can calculate that the recession of the Moon amounts to one-third of an inch per rotation. Thus, each time you see a new moon it is that much farther away from you. One-third of an inch per month is a tiny change as astronomical distances go, but, on the other hand, the Earth-Moon system must have existed for billions of years. Putting these figures together, George Darwin found that between four and five billion years ago the Earth and the Moon must have been very close together, and he suggested that they may once have been a single body (Earthoon or Moorth). The breakup into two parts may have been caused by the tidal force of solar gravity or by some other catastrophic event lost in the long ago of the solar system. Darwin's hypothesis is a source of violent disagreement among the scientists interested in the origin of the Moon. While some are ardent believers (if only because of its beauty), others are bitter enemies.

A few more words may be said about the future of the Moon as it can be calculated on the basis of celestial mechanics. As a result of gradual recession, the Moon eventually will get so far from the Earth that it will become rather useless as a substitute for lanterns at night. In the meantime solar tides gradually will slow down the rotation of the Earth (provided the oceans do not freeze up), and there will come the time when *the length of a day will be greater than the length of a month*. The friction of lunar tides then will tend to accelerate the rotation of the Earth, and, by the law of conservation of angular momentum, the Moon will begin to return to the Earth until at last it will come as close to the Earth as it was at birth. At this point the Earth's gravity forces will probably tear up the Moon into a billion pieces, forming a ring similar to that of Saturn. But the dates of these events, as given by celestial mechanics, are so far off that the Sun probably will have run out of its nuclear fuel and the entire planetary system will be submerged in darkness.

Chapter 7

TRIUMPHS OF CELESTIAL MECHANICS

Within a century the seed planted by Newton's formulation of the Law of Universal Gravity and his invention of calculus grew into a beautiful but dense forest. In the calculations of the great French mathematicians, such as Joseph Louis Legrange (1736–1813) and Pierre Simon Laplace (1749–1827), celestial mechanics reached a perfection never achieved before in science. Starting from the simplicity of Kepler's laws of planetary motion, which would have been exact if the planets moved exclusively under the action of solar gravity, the theory progressed to a high degree of complexity by taking into account the mutual interactions or *perturbations* between the planets. Of course, since planetary masses are much smaller than the mass of the Sun, the perturbations of their motions because of mutual gravitational interaction are very small, but it must not be ignored if exactness comparable to that of the precise astronomical measurements is to be attained. These kinds of calculations take a tremendous amount of time and labor (eased up today by the use of electronic computers). For example, an American astronomer, E. W. Brown, spent about two decades studying several thousand terms in the long mathematical series for computing data for his three volumes in quarto, *Tables of the Moon.*

But these laborious studies quite often brought fruitful results. Near the middle of the last century a young French astronomer, J. J. Leverrier, while comparing his calculations of the motion of the planet Uranus, accidentally discovered in 1781 by William Herschel, with its observed positions in the sixty-three years since its discovery, found that there must be something wrong. The discrepancies between the observations and calculations were as annoyingly high as 20 angular seconds (the angle subtended by a man 10 miles away), and this difference was beyond any possible error of either observation or theory. Leverrier suspected that the discrepancies were due to the perturbations caused by some unknown planet moving outside the orbit of Uranus, and he sat down to calculate how massive this hypothetical planet must be and how it would have to move to fit the observed deviations in the motion of Uranus. In the fall of 1846 Leverrier wrote to J. G. Galle, at the Berlin Observatory: "Direct your telescope to the point on the ecliptic in the constellation of Aquarius, in longitude 326°, and you will find within a degree of that place a new planet, looking like a star of about the 9th magnitude, and having a perceptible disc."

Galle followed the instructions. The new planet, which was called Neptune, was found on the night of September 23, 1846. An Englishman, J. C. Adams, fairly shares with Leverrier the honor for mathematical discovery of Neptune, but T. Challis at the University of Cambridge Observatory, to whom Adams communicated his results, was too slow in the search and thus missed the boat.

The story repeated itself, in less dramatic form, in the first half of the present century. The American astronomers W. H. Pickering, of Harvard Observatory, and Percival Lowell, the founder of Lowell Observatory in Arizona, were arguing in the late twenties that the perturbations of the motions of Uranus and Neptune suggested the existence of still another planet beyond Neptune. But it took more than ten years until this planet, which is called Pluto and may be an escaped satellite of Neptune, was actually found in 1930 by C. W. Tombaugh, of the Lowell Observatory. It seems to be a matter of opinion whether this discovery was actually due to prediction or to laborious systematic search.

Another interesting example of the exactness of the results of celestial mechanics is the use of the calculations of the dates of solar and lunar eclipses to establish historical references here on Earth. In 1887 the Austrian astronomer Theodore von Oppolzer published tables containing calculated data of all past solar and lunar eclipses, beginning with 1207 B.C., and all future ones up to A.D. 2162—altogether about 8000 solar and 5200 lunar eclipses. Using this data, one finds, for example, that we are four years behind in our calendar. Indeed, according to historical records, the Moon went into eclipse as a means of "mourning the death" of the Judean King Herod, who in the last year of his reign ordered the massacre of all children in the city of Bethlehem, hoping that the baby Christ would be among them. According to von Oppolzer's tables, the only lunar eclipse which fits the facts occurred on March 13 (Friday?) 3 B.C., and we are led to conclude that Jesus Christ was born four years earlier than our customary calendar indicates.

Other examples of historically important eclipses are that of April 6, 648 B.C., which permits us to fix with certainty the earliest date in Greek chronology, and the eclipse of 911 B.C., which establishes the chronology of ancient Assyria.

Of peculiar interest for us, the inhabitants of the Earth, is the calculation of the perturbations of the

Earth's orbit by other planets. The ellipse along which the Earth moves around the Sun does not remain invariant, as it would if the Earth were a single planet, but slowly wobbles and pulsates under the gravity forces of the other members of the solar system. We have seen in Chapter 5 that lunar-solar precession makes the rotation axis of our globe describe a conical surface in space with the period of 25,800 years. In addition, the orbit of the Earth is slowly changing its eccentricity and its tilt in space under the action of gravitational forces exerted by other planets of the solar system. The resultant changes can be calculated with great precision by the methods of celestial mechanics; they are shown in Fig. 18 for 100,000 years in the past and 100,000 years in the future. The upper part of this figure gives the changes of the eccentricity of the Earth's orbit and the rotation of its major axis. The orbit of the Earth, though elliptical, differs very little from a circle, so that its focus is very close to the geometrical center of the ellipse. The traveling white circle represents the motion of the focus in respect to the center of the orbit (large black dot). When the two points are far from each other, the eccentricity of the orbit is large; when they are close, the eccentricity is small, and if the two points coincided, the ellipse would become a circle. In the scale of this diagram, the diameter of the orbit itself would be about thirty inches.

The lower figure gives the change of the tilt of the orbit in respect to the invariant plane in space. What is plotted here is the motion of the intersection point of a perpendicular to the plane of the orbit with the sphere of the fixed stars. We notice that 80,000 years ago the eccentricity of the Earth's orbit was fairly high and that it is much smaller now (crossed circle), and will become still smaller in 20,000 years.

The changes of the Earth's orbit have a profound

effect on the climate of our globe. The increase of eccentricity changes the ratio between the least and greatest distance from the Sun, which increases the

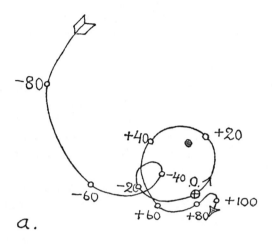

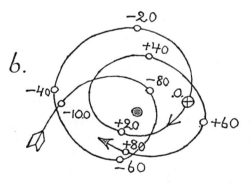

Fig. 18. Changes in the eccentricity (a) and inclination (b) of the Earth's orbit, caused by planetary perturbations. Figures show thousands of years in the past or the future.

difference in summer and winter temperatures. The increase of inclination of the Earth's axis to the plane of its orbit also increases summer and winter differences since, indeed, we know that if the rotation axis of the Earth were perpendicular to its orbit, Earth's temperature would be constant all the year around. The Serbian astronomer M. Milankovitch attempted, in 1938, to use these differences to explain glacial periods during which sheets of ice from the north periodically advanced and retreated over the lowland in middle latitudes. Milankovitch followed the calculations of Leverrier, similar to those presented in Fig. 18 but extending 600,000 years back in time. For his standard Milankovitch took the amount of solar heat now falling during the summer months on a unit surface at 65° northern latitude, and calculated for various past eras how far north or south one would have to go to find the same amount of heat. The results of these calculations are shown in Fig. 19a, overlapped on the contour of the northern shores of Eurasia. Large maxima indicate an essential decrease in solar heat, while the minima indicate the increases. Thus, for example, a little over 100,000 years ago the amount of heat arriving at the latitude of 65° north (central Norway) was comparable with what is arriving today at the latitude of Spitzbergen. On the other hand, only about 10,000 years ago central Norway enjoyed the present solar climate of Oslo and Stockholm. The curve in Fig. 19b represents the southward advance of the ice sheets, as indicated in geological data, and we notice that the agreement between the two curves is striking indeed.

The curve in Fig. 19c, corresponding only to the last 100,000 years, was published in 1956 by Hans Suess, of the University of California, and represents the temperature of the ocean waters during the past geological eras, estimated by an ingenious method first proposed in 1951, by the famous American scientist Harold Urey. This method is based on the fact that the ratio of heavy and light isotopes of oxygen

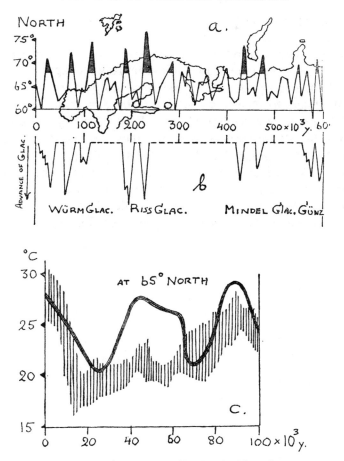

Fig. 19. Comparisons of Milankovitch's climatic curves (a) with the past advances of glaciers (b) and with paleotemperatures of the ocean (c).

(O^{18} and O^{16}) in the sedimentary deposits of calcium carbonate ($CaCO_3$) at the ocean bottom depends on the ocean water temperature at the period of sedimentation. Thus, measuring the O^{18}/O^{16} ratio in the deposits at different depths below the ocean floor, one can tell the temperature of the water one hundred thousand years ago with the same certainty that one can measure it on a thermometer lowered from a ship. Suess's temperature curve of ocean waters for the past 100,000 years stands in a reasonably good agreement with the part of Milankovitch's temperature curve calculated for the same period. Thus, in spite of the objection of some climatologists that "a few degrees difference in temperature could not have produced glacial periods," it seems that the old Serb was right after all. Therefore, we should conclude that although planets do not affect the lives of individual persons (as astrologists insist), they certainly do affect the life of Man, animals, and plants in the long run of geological history.

Chapter 8

ESCAPING GRAVITY

"What goes up must come down," is a classical saying which is not true any longer. Some of the rockets shot in recent years from the surface of the Earth have become artificial satellites of the Earth, with indefinitely long lifetimes, while others have been forever lost in the vast expanse of interplanetary space. Using the notion of gravitational potential explained in Chapter 4, we can easily calculate the velocity with which an object must be hurled up from the surface of the Earth if it is never to come back. We have seen that the work to be done in lifting a mass m from the surface of the Earth to the distance R from its center is:

$$GMm \left(\frac{1}{R_0} - \frac{1}{R} \right)$$

where G is the gravitational constant, M the mass of the Earth, m the mass of the object, and R_0 the radius of the Earth. If the object is to go beyond the point of return, we must put $R = \infty$ (infinity), $\frac{1}{R} = 0$. Thus, the work done in this case becomes:

$$\frac{GMm}{R_0}$$

On the other hand, kinetic energy of an object with

the mass m moving with the velocity v is

$$\frac{1}{2}\,mv^2$$

Thus, in order to communicate to it a sufficient amount of energy to overcome the forces of terrestrial gravity, one must satisfy the condition:

$$\frac{1}{2}\,mv^2 \geq \frac{GMm}{R_0}$$

with the symbol $\geq$ meaning "equal to" or "greater than." Since m cancels from both sides of that equation, we conclude that: *it takes the same velocity to throw an object out of the reach of the Earth's gravity no matter whether it is a light or a heavy object.*

From the above equation we obtain:

$$v \geq \sqrt{\frac{2GM}{R_0}}$$

and, putting $R_0 = 6.37 \cdot 10^8$ cm; $M = 6.97 \cdot 10^{27}$ gm, and $G = 6.66 \cdot 10^{-8}$, we find for the velocity $11.2\,\frac{km}{sec} = 25,000\,\frac{miles}{hour}$ This is the so-called *escape-velocity*, the minimum velocity at which the object will not fall back.

The situation is complicated, of course, by the presence of the Earth's atmosphere. If one had shot an artillery projectile with the necessary escape velocity from the surface of the Earth, as was described in *The Journey around the Moon*, a fantasy by the famous French science-fiction writer Jules Verne, the shell would never have arrived. Contrary to Jules Verne's description, such a projectile would have melted right away, from heat developed by air friction, and the debris would have fallen down, having lost all initial energy. Here is where the advantage of a rocket over an artillery shell comes in. A rocket starts from its launching pad quite slowly and gradually gains velocity as it climbs. Thus, it passes through the dense layers of the terrestrial atmosphere with velocities at which the friction-heating is not yet important, and gets its full speed at heights where the air is too rare to present any significant resistance to flight. Of course, the air friction in the beginning of the flight does result in some losses of energy, but these losses are comparatively small.

We can now survey what happens when a rocket, having passed through the terrestrial atmosphere and having burned all its propelling fuel, begins its journey through space. In Fig. 20 we give a graphic presentation of the gravitational potential in the region of the inner planets of the solar system (Mercury, Venus, Earth, and Mars). The main slope is due to the gravitational attraction of the Sun given by $GM\odot/r$ where $M\odot$ is the mass of the Sun, and r the distance of the rocket from it. On this general slope are overlapped local "gravitational dips" caused by the attraction of individual planets. The depths of dip are indicated in the correct scale, but their widths are strongly exaggerated, since otherwise they would look on the drawing just like vertical lines. In the lower right corner of the diagram is shown (on a much larger scale) the distribution of gravitational potential in the space between the Earth and the Moon. Since Earth-to-Moon distance is much smaller than the Earth-to-Sun distance, the change of the solar gravitational potential in this region is practically unnoticeable. Thus, in order to send a rocket to the Moon, one has only to overcome terrestrial gravity and to have sufficient velocity left to cover the distance within a reasonable time. In October 1959 Russian rocketeers accomplished this feat and managed to photograph the opposite side of the Moon. Fig. 21 gives the trajectory of that rocket, called Lunik, on its way to the Moon and back.

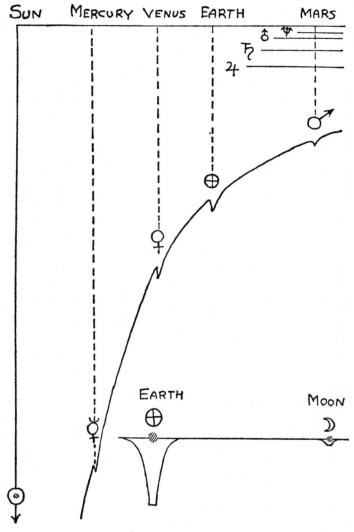

Fig. 20. Gravitational potential slope in the neighborhood of the Sun. Down on the right Earth-Moon gravitation potential.

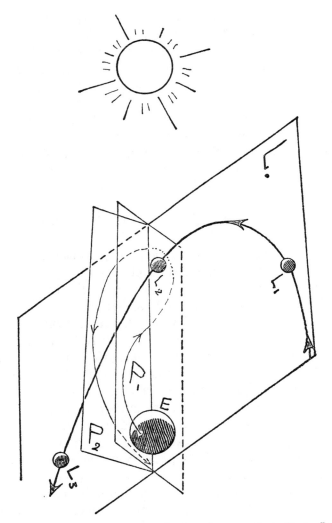

Fig. 21. The trajectory of the first rocket that flew around the Moon.

The rockets aimed at other planets of the solar system will have to cope not only with the gravitational pull of the Earth but also with the pull exerted by the Sun.

When a rocket escapes the Earth's gravity with a small leftover velocity, it is bound to move closely along the Earth's orbit, not coming any closer to or any farther away from the Sun. To get away from the Earth's orbit, the rocket must have enough velocity to climb the slope of the solar gravitational curve. As can be seen in Fig. 20, the height to be climbed in order to reach the orbit of Mars is about 6.5 times greater than the depth of the Earth's gravitational pit. Since the kinetic energy of motion increases as the square of the velocity, such a rocket must have at least the velocity of

$$11.2 \times \sqrt{6.5} = 28 \frac{\text{km}}{\text{sec}}$$

Why not choose an easier job and go down to Venus rather than up to Mars? Ironically enough, for ballistic missiles it is just as difficult to go down the slopes as to go up the slopes. The point is that the rocket, after escaping the Earth's gravity, will be bound to the Earth's orbit. If the rocket is to get farther away from the Sun, its speed must be considerably increased, which would require large additional amounts of fuel. But to come closer to the Sun is not much easier! Since a rocket coasting through empty space cannot put on the brakes to reduce its speed, as a car can, its velocity can be decreased only if the rocket ejects a powerful jet from the front, which would require about the same amount of fuel as speeding up by ejecting a jet from the rear. But, since the orbit of Venus is closer to us than that of Mars, the difference of the gravitational potential is only five times as large as the gravitational dip of the Earth, and the task is correspondingly easier. And, in fact, on the 12th of February, 1961, Russian rocketeers dispatched a rocket toward Venus. It never came back.

All the rockets so far sent into space have been

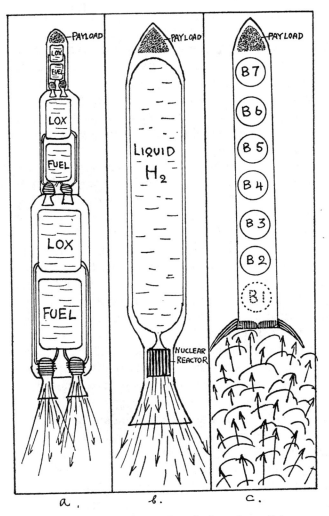

Fig. 22. (a) A multistage chemical rocket; (b) conventional nuclear rocket; (c) unconventional nuclear rocket.

propelled by ordinary chemical fuel and based on the multistage principle illustrated in Fig. 22a. Several rockets of decreasing sizes are arranged one atop another, and the journey is started by firing the motors of the first stage, i.e., of the largest rocket at the bottom. When this modern totem pole reaches the maximum upward velocity and the fuel tanks of the first stage rocket are empty, it is separated from the rest and the second stage rocket's motors are started. The process is continued until the last stage, containing the instrumentation, mice, monkeys, or men, is finally accelerated to the needed velocity.

Another possibility under intensive study at the present time is the use of nuclear energy. It must be remembered that the propulsion of space ships presents entirely different problems from the propulsion of sea or air ships. For the latter all we need is energy, since those ships move forward by pushing against the surrounding medium, be it water or air. One cannot push against a vacuum, and space ships are propelled by ejecting through nozzles some material which the ships carry. In the ordinary chemical-fuel rockets we have a two-in-one situation. Energy is produced by chemical reaction between the fuel and the oxidizer carried in two separate tanks, and the products of that reaction serve as the material ejected from the nozzle. The advantage of using the products of the energy-producing process as the ejected material is counterbalanced, however, by the fact that the products of burning (mostly carbon dioxide and water vapor) consist of comparatively heavy molecules. It is shown by the theory of jet-driven vehicles that the thrust decreases with the increasing weight of the molecules forming the jet. Thus, it would be advantageous to use for jets the lightest chemical element, hydrogen, but of course hydrogen, being an element, is not produced as a result of any kind of burning. What might be done, however, would be to carry liquid hydrogen in a single tank and heat it up to very high temperature with some kind of nuclear reactor. A schematic drawing of such a nuclear rocket is shown in Fig. 22b.

Another promising proposal for using nuclear energy for rocket propulsion, originally advanced by Dr. Stanislaw Ulam, of Los Alamos Scientific Laboratory, is shown in Fig. 22c. The body of the rocket is filled with a large number of small atomic bombs, which are ejected one by one from the opening in the rear and exploded some distance behind the rocket. The high-velocity gases from these explosions would overtake the rocket and exert pressure on a large disc attached to its rear. These successive kicks would speed up the rocket until it reached the desired velocity. The preliminary studies of such a method of propulsion indicate that it might be superior to a reactor-heated hydrogen design.

It is difficult, in a nontechnical book such as this one, to describe all the possibilities dawning on the horizons of space flight progress, and we conclude this chapter by stressing one important point. In sending space ships to the faraway points of our solar system (and maybe beyond) one faces two distinctly different problems: First, how to escape from the gravitational pull of the Earth? Second, how, after escaping, to get enough velocity to travel to our destinations? So far all attempts in this direction have been limited to the task of giving a rocket enough initial speed to escape terrestrial gravity with enough velocity left over to proceed elsewhere. One can, however, separate these two tasks and use different propulsion methods for the first and for the second step.

To get away from the Earth's surface requires a violent action since, if the thrust of the rocket motors is not great enough, the rocket will huff and puff but not lift itself from its launching pad. Here powerful chemical or nuclear propulsion methods are necessary. Once the space ship is lifted and put on a satellite orbit around the Earth, the situation becomes quite different. We now have plenty of time to accelerate the space ship and can use less violent and more economical methods of propulsion. It still can be chemical or nuclear energy or, for that matter, the energy supplied by the Sun's rays, but one is not in a hurry and not in danger of falling down. A space ship put into orbit around our globe can take time to accelerate its flight and, moving along a slowly unwinding spiral trajectory, finally muster enough speed to accomplish its task. It is very likely that the combination of violent action at the start and a more leisurely sailing for the rest of the trip will be the future solution of the problem of space travel.

Chapter 9

EINSTEIN'S THEORY OF GRAVITY*

* The content of this and the next chapter follows closely the author's article "Gravity" published in the March 1961 issue of *Scientific American*.

The tremendous success of Newton's theory in describing the motions of celestial bodies down to their most minute details characterized a memorable era in the history of physics and astronomy. However, the *nature* of gravitational interaction and, in particular, the reason for the proportionality between gravitational and inertial mass, which makes all bodies fall with the same acceleration, remained in complete darkness until Albert Einstein, in 1914, published a paper on the subject. A decade earlier Einstein had formulated his Special Theory of Relativity, in which he postulated that no observation made inside an enclosed chamber, even if one could turn the chamber into a most elaborate physical laboratory, would answer the question whether the chamber was at rest or moving along a straight line with constant velocity. On this basis Einstein rejected the idea of absolute uniform motion, threw out the ancient and contradictory notion of "world ether," and erected his Theory of Relativity, which revolutionized physics. Indeed, no mechanical, optical, or any other physical measurement one could make in an inside cabin of a ship sailing a smooth sea (this chapter is being written in an inside cabin of the S.S. *Queen Elizabeth*) or in an airplane flying through quiet air with the window curtains drawn, could possibly give any information as to whether the ship was afloat or in dry dock, the plane airborne or at the airport. But, if the sea is choppy or the air is rough, or if the ship hits an iceberg or the plane a mountaintop, the situation becomes entirely different; any deviation from uniform motion will be painfully noticeable.

To deal with this problem Einstein imagined himself in the position of a modern astronaut and considered what would be the results of various physical experiments in a space observation station far from any large gravitating masses (Fig. 23). In such a station at rest or in uniform motion in respect to distant stars, the observers inside the laboratory, and all the instruments not secured to the walls, would float freely within the chamber. There would be no "up" and no "down." But, as soon as the rocket motors were started, and the chamber accelerated in a certain direction, phenomena very similar to gravity would be observed. All the instruments and people would be pressed to the wall adjacent to the rocket motors. This wall would become the "floor" while the opposite wall became the "ceiling." The people would be able to rise on their feet and stand very much as they stand on the ground. If, furthermore, the acceleration of the space ship was made equal to the acceleration of gravity on the surface of the Earth, the people inside could well believe that their ship was still standing on its launching pad.

Suppose that, in order to test the properties of this "pseudogravity," an observer within an accelerated rocket should release simultaneously two spheres, one of iron and one of wood. What "actually" would happen can be described in the following words: While the observer holds the two spheres in his hands, the spheres are moving in an accelerated way, along with the rocket ship driven by its motors. As soon as he releases the spheres, however, and thus disconnects them from the rocket's body, no driving force will act on them any

Fig. 23. Albert Einstein in an imaginary (*gedanken-experimental*) rocket.

more, and the spheres will move side by side with a velocity equal to that of the space ship at the moment of release. The rocket ship itself, however, will continuously gain speed, and the "floor" of the space lab will quickly overtake the two spheres and "hit" them simultaneously. To the observer who has released the two balls the phenomenon will seem otherwise. He will see the spheres fall and "hit the floor" at the same time. And he will remember Galileo's demonstration on the Tower of Pisa, and will become still more persuaded that a regular gravitational field does exist in his space laboratory.

Both descriptions of what the spheres would do are equally correct, and Einstein incorporated the equivalence of the two points of view in the foundation of his new, relativistic theory of gravity. This so-called *principle of equivalence* between observations carried out within an accelerated chamber and in a "real" field of gravity would, however, be trivial if it applied only to mechanical phenomena. It was Einstein's idea that this equivalence is quite general and holds also in the case of optical and all electromagnetic phenomena.

Let us consider what happens to a beam of light propagating across our space chamber from one wall to the other. We can observe the path of light if we put a series of fluorescent glass plates across it or simply if we blow cigarette smoke into the beam. Fig. 24 shows what "actually" happens when a beam goes through several glass plates placed at equal distance from one another. In (a) the light hits the upper section of the first plate, producing a fluorescent spot. In (b) when the light reaches the second plate, it produces fluorescence closer to the middle of the plate. In (c) the light hits the third plate still lower. Since the motion of the rocket is accelerated, the distance traveled during the second time interval is three times greater than during the first one, and, hence, the three fluorescent spots will not be on a straight line but on a curve (parabola) bent downward. The observer inside the chamber, considering all the phenomena he observes as due to gravity, will conclude from his experiment that *the light ray is bent when propagating through a gravitational field.* Thus, concluded Einstein, if the principle of equivalence is a general principle of physics, light rays from distant stars

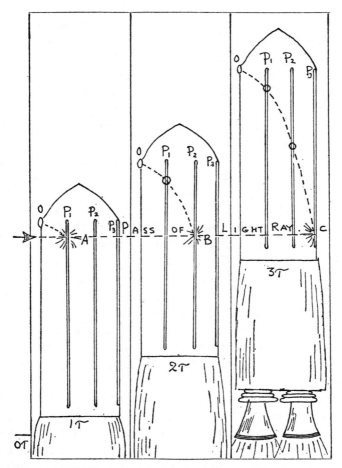

Fig. 24. Light propagation in an accelerated rocket.

should be bent if they pass close to the surface of the Sun on the way to a terrestrial observer. His conclusion was brilliantly confirmed in the eclipse of 1919 when a British astronomical expedition to Africa observed the displacement of the apparent positions of stars in the neighborhood of the eclipsed Sun. Thus the equivalence of the gravitational field and the accelerated systems became an indisputable fact of physics.

We shall turn now to another type of accelerated motion and its relation to the gravitational field. So far we have talked about the case when the velocity changes its numerical value but not its direction. There also is the type of motion in which the velocity changes its direction but not its numerical value—i.e., rotational motion. Imagine a merry-go-round (Fig. 25) with a curtain hanging all around it so that the people inside cannot tell by looking at the surroundings that the platform is rotating. As everybody knows, a person standing on a rotating platform seems to be subject to centrifugal force, which pushes him toward the rim of the platform, and a ball placed on the platform will roll away from the center. Centrifugal force acting on any object placed on the platform is proportional to the object's mass, so here again we can consider things as being equivalent to the field of gravity. But it is a very peculiar gravitational field, and rather different from the fields surrounding the Earth or the Sun. First of all, instead of representing the attraction, which decreases as the square of the distance from the center, it corresponds to a repulsion increasing proportionally to that distance. Secondly, instead of being spherically symmetrical around the central mass, it possesses a cylindrical symmetry around the central axis, which coincides with the rotation axis of the platform. But Einstein's equivalence principle works here, too, and those forces can be interpreted as being caused by gravitating masses distributed at large distances all around the symmetry axis.

Physical events occurring on such a rotating platform can be interpreted on the basis of Einstein's Special

Theory of Relativity, according to which the length of measuring rods and the rate of clocks are affected by their motion. Indeed, the two basic conclusions of that theory are:

1. If we observe an object moving past us with a certain velocity v it will look contracted in the direction of its motion by a factor

$$\sqrt{1 - \frac{v^2}{c^2}}$$

where c is the velocity of light. For ordinary speeds, which are very small as compared to the velocity of light, this factor is practically equal to one, and no noticeable contraction will be observed. But when v approaches c, the effect becomes of great importance.

2. If we observe a clock moving past us with the velocity v it will appear to be losing time, and its rate will be slowed down by a factor

$$1 \Big/ \sqrt{1 - \frac{v^2}{c^2}}$$

As in the case of spacial contraction, this effect can be observed only when the velocity v is approaching that of light.

Keeping in mind these two effects, let us consider the results of various observations which can be made on a rotating platform. Suppose we want to find the laws of propagation of light between different points on the platform. We select two points, A and B, on the periphery of the rotating disc (Fig. 25a), one serving as the source and another as the receptor of light. According to the basic law of optics, light always propagates along the shortest path. What is the shortest path between the points A and B on the rotating platform? To measure the length of any line connecting A and B,

GRAVITY

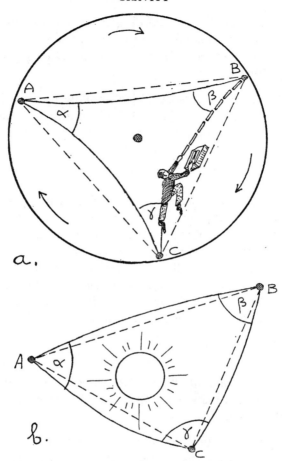

Fig. 25. (a) Some experiments on a rotating platform; (b) triangulation around the Sun.

we will use here an old-fashioned but always safe method of counting the number of yardsticks which can be placed end to end along the line between *A* and *B*. If the disc is not rotating, the situation is obvious and the shortest distance between *A* and *B* is along the straight line of good old Euclidean geometry. But if the disc is rotating, the yardsticks placed along the *AB* line are moving with a certain velocity, and are therefore expected to undergo the relativistic contraction of their length. One then will need a larger number of sticks to cover that distance. Here, however, an interesting situation crops up. If one moves a yardstick closer to the center, its linear velocity becomes smaller and it will not contract so much as when it was farther away. Thus, bending the line of yardsticks toward the center, we will need a lesser number of yardsticks since, although the "actual" distance is somewhat longer, this disadvantage will be overcompensated by lesser shrinkage of each yardstick. If we substitute light waves for yardsticks, we come to a conclusion that the light ray too will be bent in the direction of the gravitational field, which here is directed away from the center.

Before leaving our merry-go-round platform let us perform one more experiment. Let us take a pair of identical clocks, place one in the center of a platform and another at its periphery. Since the first clock is at rest while the second is moving with a certain velocity, the second will lose time in respect to the first one. Interpreting centrifugal force as a force of gravity, one will say that the clock placed in the higher gravitational potential (that is, in the direction in which gravitational force acts) will move more slowly. This slowing down will apply equally to all other physical, chemical, and biological phenomena. A typist working on the first floor of the Empire State Building will age more slowly than her twin sister working on the top floor. The difference will be very small, however; it can be calculated that in ten years the girl on the first floor will be a few millionths of a second younger than her twin on the top floor. In the difference in gravity between the surface of the Earth and the surface of the Sun, the effect is considerably larger. A clock placed on the surface of the

Sun would slow down by one ten-thousandth of a per cent in respect to the terrestrial clock. Of course, nobody can place a clock on the surface of the Sun and watch it go, but the expected slowing down was confirmed by observing the frequencies of spectral lines emitted by atoms in the solar atmosphere.

The problem of twin sisters' aging at a different rate because they work in places having different gravitational potentials is closely related to the problem of twin brothers, one of whom sits home while the other travels a lot. Let us imagine twin brothers, one a spaceship pilot and another an employee at the space terminal somewhere on the surface of the Earth. The pilot brother starts on a mission to some distant star, flying his space ship with a velocity close to that of light, while his twin brother continues his office work at the space terminal. According to Einstein, each brother ages more slowly than the other. Thus, when the pilot brother returns to the Earth, one is led to expect that he will find that his office brother has aged less than himself, but the office brother will come to an exactly opposite conclusion. This is apparent nonsense since, for example, if one measures age by graying of the hair, the two brothers could find out who had aged more, simply by standing side-by-side in front of a mirror.

The answer to this alleged paradox is that the statement concerning the relative aging of the twin brothers is correct *only* within the frame of the so-called Special Theory of Relativity, which considers only uniform motion at constant velocity. In this case the pilot brother will certainly never come back and therefore cannot possibly stand side-by-side with his office brother in front of the mirror to compare graying hair. The best the brothers can do is to have two TV sets: one in the terminal office showing the pilot brother and his clock in the cockpit of the space ship, the other in the space ship, showing the office brother at his desk and the office clock above his head (Fig. 26).

Dr. Eugene Feenberg, of Washington University, investigated this situation theoretically on the basis of

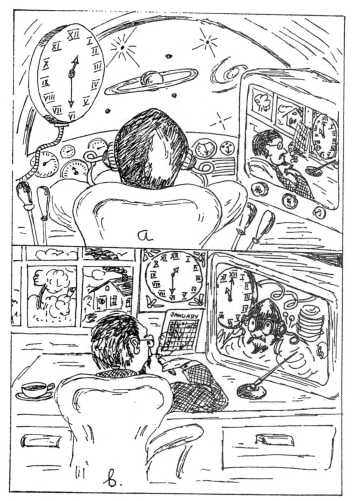

Fig. 26. Relative aging of twin brothers, as observed on TV sets.

the well-known laws of the propagation of radio signals, and the conclusion was that looking at the TV screen, *each* brother indeed will observe that the other ages more slowly. But if the flying brother has to come back, he must first decelerate his space ship, bring it to a complete stop, and accelerate it homeward. This necessity puts the twin brothers entirely in different positions. As we have seen before, acceleration and deceleration are equivalent to a gravitational field which slows down the rate of the clock, as well as the rates of all other phenomena. And, just as a typist working on the first floor of the Empire State Building will age more slowly than her twin sister working on the top floor, the flying brother will age more slowly than his twin brother on the ground. Thus, if the flight is long enough, the returning pilot will twirl his black mustache as he looks at the shining bald head of his twin. Therefore, there is no paradox here at all.

An interesting experiment designed to confirm the slowing down of time by gravity (if further confirmation is necessary) is proposed by S. F. Singer of the University of Maryland, who suggests placing an atomic clock in the satellites traveling along circular orbits at different altitudes above the Earth's surface. It was calculated that for a satellite traveling at altitudes less than the radius of our globe, the main relativistic effect will be that of slowing down the clock as a result of its velocity and will be given by the time-dilating factor $\sqrt{1 - \dfrac{v^2}{c}}$. For higher altitudes, however, the velocity effect is expected to become of smaller importance and, instead of losing time, the clock will gain time because it is in the weaker gravitational field (as would be the girl working on the top of the Empire State Building). There is hardly any doubt that this interesting experiment will confirm Einstein's theory.

This discussion brings us to the conclusion that light, propagating through a gravitational field, does not follow a straight line but curves in the direction of the field and that, due to the shrinkage of yardsticks, the shortest distance between two points is not a straight line but a curve also bending in the direction of the gravitational field. But, what other definition can one give to a "straight line" than the path of light in vacuum or the shortest distance between two points? Einstein's idea was that one should retain the old definition of a "straight line" in the case of the gravitational field, but instead of saying that light rays and shortest distances are curved, say that the space itself is curved. It is difficult to conceive the idea of a curved three-dimensional space, and even more so of a curved four-dimensional space in which time serves as the fourth coordinate. The best way is to use an analogy with the two-dimensional surfaces which we can easily visualize. We are all familiar with plane Euclidean geometry, which pertains to the various figures you can draw on a flat surface, or plane. But if, instead of a plane, we draw geometrical figures on a curved surface, such as a surface of a sphere, the Euclidean theorems no longer hold. This is demonstrated in Fig. 27, which represents triangles drawn on a plane (a), on the spherical surface (b), and on a surface which (for obvious reasons) is called a saddle-surface (c).

For a plane triangle the sum of three angles is always equal to 180°. For a triangle on the surface of the sphere the sum of the three angles is always larger than 180° and the excess depends on the ratio of the size of the triangle to the size of the sphere. For triangles drawn on a saddle-surface, the sum of the angles is less than 180°. True enough, the lines forming triangles on spherical and saddle-surfaces are not "straight" from the three-dimensional point of view, but they are the "straightest"—i.e., the shortest—distances between the two points if one is confined to the surface in question. Not to confuse the terminology, mathematicians call these lines *geodesic lines* or simply *geodesics*.

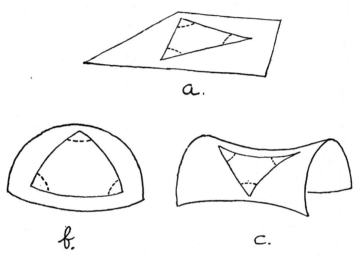

Fig. 27. Triangles on a plane surface (a), a sphere (b), and a saddle-surface (c).

Similarly, we can speak about geodesical or shortest lines in three-dimensional space connecting two points along which light rays would propagate. And, measuring the sum of the three angles of a triangle in space, we can call the space flat if this sum is equal to 180°, sphere-like or positively-curved if this sum is larger than 180°, and saddle-like or negatively-curved if it is less than 180°. Imagine three astronomers on Earth, Venus, and Mars measuring the angles of a triangle formed by light rays traveling between these three planets. Since, as we have seen, light rays propagating through the gravitational field of the Sun bend in the direction of the force of gravity, the situation will look as shown in Fig. 25b, and the sum of the angles of the triangle will be found to be larger than 180°. It would be reasonable to state in this case that light propagates along the shortest distances, or geodesical lines, but that the space around the Sun is curved in the positive sense. Similarly, in the gravitational field, which is equivalent to the field of centrifugal force on a rotating disc (Fig. 25a), the sum of angles of a triangle is smaller than 180°, and that space must be considered to be curved in the negative sense.

The foregoing arguments represent the foundation of Einstein's geometrical theory of gravity. His theory supplanted the old Newtonian point of view, according to which large masses such as the Sun produce in the surrounding space certain fields of forces which make planets move along the curved trajectories instead of straight lines. In the Einsteinian picture the space itself becomes curved while the planets move along the "straightest"—i.e., geodesical—lines in that curved space. It should be added, to avoid a misunderstanding, that we refer here to geodesical lines in the four-dimensional space-time continuum, and that it would be wrong, of course, to say that the orbits themselves are geodesical lines in the three-dimensional space. The situation is schematically illustrated in Fig. 28, which shows the time axis, t, and two space axes, x and y, lying in the plane of the orbit. The winding line, known as the *world-line* of a moving object (in this case the Earth), *is* the geodesic line in the space-time continuum.* Einstein's interpretation of gravity as the curvature of the space-time continuum leads to results slightly different from the prediction of the classical Newtonian theory, thus permitting observational verifi-

* It must be noticed here that the vertical and horizontal scales in Fig. 28 are given by necessity in different units. Indeed, while the radius of the Earth's orbit is only 8 minutes (if expressed in the time of light propagation), the distance from one January plane to another is, of course, one year—i.e., sixty thousand times longer. Thus, in proper scale the geodesic line will indeed deviate from a straight line but very little.

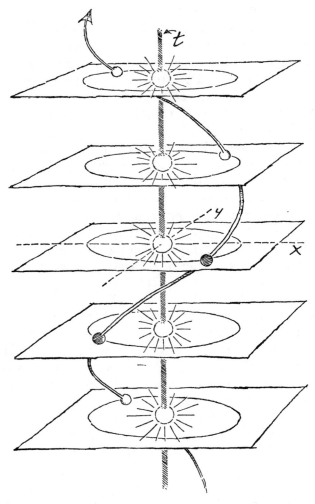

Fig. 28.   World line of the moving Earth in the space-time continuum is represented here in a co-ordinate system with the vertical time axis *t* and two space axes *x* and *y*.

cation. For example, it explained the precession of the major axis of Mercury's orbit by 43 angular seconds per century, and so solved a long-standing mystery of classical celestial mechanics.

Chapter 10

UNSOLVED PROBLEMS OF GRAVITY

In the laboratory diary of Michael Faraday (1791–1867), who made many important contributions to the knowledge of electricity and magnetism, there is an interesting entry in 1849. It reads:

> Gravity. Surely this force must be capable of an experimental relation to electricity, magnetism, and other forces, so as to build it up with them in reciprocal action and equivalent effect. Consider for a moment how to set about touching this matter by facts and trial.

But the numerous experiments this famous British physicist undertook to discover such a relation were fruitless, and he concluded this section of his diary with these words:

> Here end my trials for the present. The results are negative. They do not shake my strong feeling of the existence of a relation between gravity and electricity, though they give no proof that such a relation exists.

It is very odd that the theory of gravity, originated by Newton and completed by Einstein, should stand now in majestic isolation, a Taj Mahal (Fig. 29) of science, having little if anything to do with the rapid developments in other branches of physics. Einstein's concept of the gravitational field grew from his Special

Fig. 29. The Temple of Gravity (the letterings on the temple are the basic equations of Einstein's relativistic theory of gravity).

Theory of Relativity, and the Special Theory was based on the theory of the *electromagnetic field* formulated in the last century by the British physicist James Clerk

Maxwell (1831–79). But in spite of many attempts, Einstein and those who have followed him have failed to establish any contact with Maxwell's electrodynamics.

Einstein's theory of gravitation was more or less contemporary with *quantum theory*, but in the forty-five years since they appeared the two theories have had quite different rates of development. Proposed by Max Planck and carried forward by the work of Niels Bohr, Louis de Broglie, Erwin Schrödinger, Werner Heisenberg, and others, quantum theory has made colossal progress and evolved into a broad discipline that explains in detail the inner structures of atoms and their nuclei. On the other hand, Einstein's theory of gravity remains to this day essentially as it was when he formulated it half a century ago. While hundreds, even thousands, of scientists study the various branches of quantum theory and apply it in many, many fields of experimental research, only a few persist in devoting their time and passion to further development in the study of gravitation. Can it be that empty space is simpler than material bodies? Or did the genius of Einstein accomplish everything that could be done about gravity in our time and so deprive a generation of the hope of further progress?

Having reduced gravity to the geometrical properties of a space-time continuum, Einstein was persuaded that the electromagnetic field must also have some purely geometrical interpretation. The *Unified Field Theory*, which grew from this conviction, had rough going, however, and Einstein died without having produced anything in field theory as simple, elegant, and convincing as his previous work. It seems now that the true relation between gravitational and electromagnetic forces is to be found only through understanding of the elementary particles, of which we hear so much nowadays, and learning why those particular particles with those particular masses and electric charges do exist in Nature.

A basic question here pertains to the relative strength of the gravitational and electromagnetic interactions between particles. Earlier in the book we derived the gravitational law which establishes the inverse square relationship between the attracting force and the distance. The French scientist Charles A. Coulomb (1736–1806) demonstrated in 1784 an analogous inverse square law for the force between electric charges.

Suppose we consider the electric and gravitational forces between two particles of 4×10^{-26} grams mass, intermediate between the masses of proton and electron, at a distance r apart. According to Coulomb's law, electrostatic force is given by e^2/r^2 where e (4.77×10^{-10} esu)* is elementary electric charge. On the other hand, according to Newton's law, gravitational interaction is given by $\dfrac{GM^2}{r^2}$ where G (6.67×10^{-8}) is the gravitational constant and M (4×10^{-26} gm) is the intermediate mass. The ratio of the two forces is $\dfrac{e^2}{GM^2}$, which is numerically equal to 10^{40}. Any theory which claims to describe the relation between electromagnetism and gravity must explain why this electric interaction between the two particles is 10^{40} times larger than the gravitational interaction. It must be kept in mind that this ratio is a pure number and remains unchanged no matter which system of units one uses for measuring various physical quantities. In theoretical formulas one often has numerical constants which can be derived in a purely mathematical way. But these

* One electrostatic unit of charge is defined as a charge which repels an equal charge placed at the distance of 1 centimeter with a force of one dyne.

numerical constants are usually small numbers such as 2π, $\frac{3}{5}$, $\frac{\pi^2}{3}$, etc. How can one derive mathematically a constant as large as 10^{40}?

More than twenty years ago a very interesting proposal in this direction was made by a celebrated British physicist, P. A. M. Dirac. He suggested that the figure 10^{40} is not at all a constant but a variable which changes with time and is connected with the age of our Universe. According to the theory of the Expanding Universe, our Universe had its origin about 5.10^9 years or 10^{17} seconds ago. Of course, a year or a second are very arbitrary units for measuring time, and one should rather select an elementary time interval which can be derived from the basic properties of matter and light. One very reasonable way of doing it would be to choose as the elementary unit of time the time interval required by light to propagate a distance equal to the diameter of an elementary particle. Since all elementary particles have diameters of about 3.10^{-13} cm, and since the velocity of light is 3.10^{10} cm/sec, this elementary time unit is

$$\frac{3.10^{-13}}{3.10^{10}} = 10^{-23} \text{ sec}$$

Dividing the present age of the Universe (10^{17} sec) by this time interval, we obtain $10^{17}/10^{-23} = 10^{40}$, which is of the same order of magnitude as the observed ratio of electrostatic and gravitational forces. Thus, said Dirac, the large ratio of electric to gravitational forces is characteristic for the present age of our Universe. When the Universe was, say, half as old as it is now, this ratio was also one-half of its present value. Since there are good reasons to assume that elementary electric charge (e) does not change in time, Dirac concluded that it is the gravitational constant (G)

which is decreasing in time, and that this decrease may be associated with the expansion of the Universe and the steady rarefaction of the material filling it.

These views of Dirac were later criticized by Edward Teller (Father of the H-bomb), who pointed out that the variation of the gravitational constant G would result in the change of temperature of the Earth's surface. Indeed, the decrease of gravity would result in the increase of the radii of planetary orbits, which (as can be shown on the basis of the laws of mechanics) would change in inverse proportion to G. The decrease would result also in the distortion of the internal equilibrium of the Sun, leading to the change of its central temperature, and of the rate of energy-producing thermonuclear reactions.

From the theory of internal structure and energy production of stars, one can show that the luminosity* L of the Sun would change as $G^{7.25}$. Since the surface temperature of the Earth varies as the fourth root of the Sun's luminosity divided by the square of the radius of the Earth's orbit, it follows that it will be proportional to $G^{2.4}$ or inversely proportional to (time)$^{2.4}$, if G varies in inverse proportion to time. Assuming for the age of the solar system the value of three billion years, which seemed to be correct at the time of his publication, Teller calculated that during the Cambrian Era (a half billion years ago) the temperature of the Earth must have been some 50° C above the boiling point of water, so that all water on our planet must have been in the form of hot vapor. Since, according to geological data, well-developed marine life existed during that period, Teller concluded that Dirac's hypothesis concerning the variability of the gravitational constant cannot be correct. During the last decade, however, the

* Luminosity of a light source is defined as the amount of light it emits per unit time.

estimates of the age of the solar system have been changed toward considerably higher values, and the correct figure may be five billion years or even more. This would bring the temperature of the primitive ocean below the boiling point of water and make the old Teller objection invalid, provided that the Trilobites and Silurian molluscs could live in very hot water. It may also help paleontological theories by increasing the rate of thermal mutations during the early stages of the evolution of life, and supplying, during the still earlier periods, very high temperatures necessary for the synthesis of nucleic acids which, along with proteins, form the essential chemical constituents of all living beings. Thus the question of variability of the gravitational constant still remains open.

Gravity and Quantum Theory

Newton's law of gravitational interaction between masses, as we have seen, is quite similar to the law of electrostatic interaction between charges, and Einstein's theory of the gravitational field has many common elements with Maxwell's theory of the electromagnetic field. So it is natural to expect that an oscillating mass should give rise to gravitational waves just as an oscillating electric charge produces electromagnetic waves. In a famous article published in 1918 Einstein indeed obtained solutions of his basic equation of general relativity that represent such gravitational disturbances propagating through space with the velocity of light. If they exist, gravitational waves must carry energy; but their intensity, or the amount of energy they transport, is extremely small. For example, the Earth, in its orbital motion around the Sun, should emit about .001 watt, which would result in its falling a millionth of a centimeter toward the Sun in a billion years! No one has yet thought of a way to detect waves so weak.

Are gravitational waves divided into discrete energy packets, or quanta, as electromagnetic waves are? This question, which is as old as the quantum theory, was finally answered two years ago by Dirac. He succeeded in quantizing the gravitational-field equation and showed that the energy of gravity quanta, or "gravitons," is equal to Planck's constant, h, times their frequency—the same expression that gives the energy of light quanta or photons. The spin of the graviton, however, is twice the spin of the photon.

Because of their weakness gravitational waves are of no importance in celestial mechanics. But might not gravitons play some role in the physics of elementary particles? These ultimate bits of matter interact in a variety of ways, by means of the emission or absorption of appropriate "field quanta." Thus electromagnetic interactions (for example the attraction of oppositely charged bodies) involve the emission or absorption of photons; presumably gravitational interactions are similarly related to gravitons. In the past few years it has become clear that the interactions of matter fall into distinct classes: (1) strong interactions, which include electromagnetic forces; (2) weak interactions such as the "beta decay" of a radioactive nucleus, in which an electron and a neutrino are emitted; (3) gravitational interactions, which are vastly weaker than the ones called "weak."

The strength of an interaction is related to the rate, or probability, of the emission or absorption of its quantum. For example, a nucleus takes about 10^{-12} second (a millionth of a billionth of a second) to emit a photon. In comparison, the beta decay of a neutron takes 12 minutes—about 10^{14} times longer. It can be calculated that the time necessary for the emission of a graviton by a nucleus is 10^{60} seconds, or 10^{53} years! This is slower than the weak interaction by a factor of 10^{58}.

Now, neutrinos are themselves particles with an extremely low probability of absorption, that is, inter-action, with other types of matter. They have no charge and no mass. As long ago as 1933 Niels Bohr inquired: "What is the difference between neutrinos and the quanta of gravitational waves?" In the so-called weak interactions neutrinos are emitted together with other particles. What about processes involving only neutrinos —say, the emission of a neutrino-antineutrino pair by an excited nucleus? No one has detected such events, but they may occur, perhaps on the same time scale as the gravitational interaction. A pair of neutrinos would furnish a spin of two, the value calculated for the graviton by Dirac. All this is, of course, the sheerest speculation, but a connection between neutrinos and gravity is an exciting theoretical possibility.

Antigravity

In one of his fantastic stories H. G. Wells describes a British inventor, Mr. Cavor, who found a material called "cavorite" which was impenetrable to the forces of gravity. Just as sheet-copper and sheet-iron can be used for shielding from electric and magnetic forces, a sheet of cavorite would shield material objects from the forces of terrestrial gravity, and any object placed above such a sheet would lose all, or at least most, of its weight. Mr. Cavor built a large spherical gondola sur-rounded on all sides by cavorite shutters, which could be closed or opened. Getting into the gondola one night when the Moon was high in the sky, he closed all the shutters facing the ground and opened all those directed toward the Moon. The closed shutters cut off the forces of terrestrial gravity and, being subjected only to the gravity forces of the Moon, the gondola flew up into space and carried Mr. Cavor to many unusual adventures on the surface of our satellite. Why is such an invention impossible, or is it impossible? There exists a profound similarity between Newton's law of Universal Gravity, Coulomb's law of the interaction of electric charges, and Sir Humphrey Gilbert's law for the interaction of magnetic poles. And, if one can shield electric and magnetic forces, why can it not also be done with gravitational forces?

To answer this question we have to consider the mechanism of electric and magnetic shielding, which is closely associated with the atomic structure of matter. Each atom or molecule is a system of positive and negative electric charges, and in metals there is present a large number of free negative electrons moving through a crystal lattice of positively-charged ions. When a piece of material is put in an electric field, the electric charges are displaced in opposite directions, and one says that the material becomes electrically polarized. The new electric field caused by this polarization is directed opposite to the original field, and the overlap of the two reduces its strength. There is a similar effect in magnetic shielding since most atoms represent tiny magnets which become oriented when the material is placed into an external magnetic field. Here again the reduction of the field strength is due to the magnetic polarization of atomic particles.

Gravitational polarization of matter, which would make possible the shielding of the forces of gravity, would require that matter be constituted of two kinds of particles: those with positive gravitational mass which would be attracted by the Earth, and those with negative gravitational mass which would be repelled. Positive and negative electric charges as well as two kinds of magnetic poles are equally abundant in nature, but particles with negative gravitational mass are as yet unknown, at least within the structure of ordinary atoms and molecules. Thus, ordinary matter cannot be gravitationally polarized, the necessary condition for the shielding of gravity forces. But what about antiparticles with which physicists have been playing during the last few decades? Could it not be that along with their opposite charges, positive electrons, negative protons, antineutrons, and other upside-down particles also have negative gravitational masses? This question seems at first sight an easy one to be answered experimentally. All one has to do is to see whether a horizontal beam of positive electrons or negative protons coming from an accelerating machine bends down or up in the gravitational field of the Earth. Since all the particles produced artificially by nuclear bombardment methods move with velocities close to that of light, the bending of a horizontal beam by the forces of terrestrial gravity (be it up or down) is extremely small, amounting to about 10^{-12} cm (nuclear diameter!) per kilometer length of the track. Of course, one could try to slow these particles down to thermal velocities, as has been done with ordinary neutrons.* In the neutron experiment a beam of fast neutrons was shot into a moderator block, and the emerging slowed-down neutrons were observed to rain down from the block with about the same speed as rain droplets fall. But the slowing down of neutrons results from collisions with the nuclei of the moderating material, and good moderators, such as carbon or heavy water, are those substances whose nuclei have low affinity for neutrons

* See *The Neutron Story* by Donald J. Hughes, Science Study Series, 1959.

and do not swallow them in a number of successive collisions. Any moderator made from ordinary matter, of course, will be a death trap for antineutrons, which will immediately be annihilated with the ordinary neutrons in the ordinary atomic nuclei. Thus, from the experimental point of view, the question about the sign of the gravitational mass of antiparticles remains open.

From the theoretical point of view the question remains open too, since we are not in possession of the theory that could predict the relation between gravitational and electromagnetic interaction. One can say, however, that if a future experiment should show that antiparticles have a negative gravitational mass, it would deliver a painful blow to the entire Einstein theory of gravity by disproving the Principle of Equivalence. In fact, if an observer inside an accelerated Einstein chamber released an apple having a negative gravitational mass, the apple would "fall upward" (in respect to the space ship), and, as observed from outside, would move with an acceleration twice that of the space ship without being subject to any outside forces. Thus we will be forced to choose between Newton's Law of Inertia and Einstein's Principle of Equivalence—a very difficult choice indeed.

INDEX

A CATALOG OF SELECTED
DOVER BOOKS
IN ALL FIELDS OF INTEREST

A CATALOG OF SELECTED DOVER
BOOKS IN ALL FIELDS OF INTEREST

CONCERNING THE SPIRITUAL IN ART, Wassily Kandinsky. Pioneering work by father of abstract art. Thoughts on color theory, nature of art. Analysis of earlier masters. 12 illustrations. 80pp. of text. 5⅜ x 8½.   23411-8

ANIMALS: 1,419 Copyright-Free Illustrations of Mammals, Birds, Fish, Insects, etc., Jim Harter (ed.). Clear wood engravings present, in extremely lifelike poses, over 1,000 species of animals. One of the most extensive pictorial sourcebooks of its kind. Captions. Index. 284pp. 9 x 12.   23766-4

CELTIC ART: The Methods of Construction, George Bain. Simple geometric techniques for making Celtic interlacements, spirals, Kells-type initials, animals, humans, etc. Over 500 illustrations. 160pp. 9 x 12. (Available in U.S. only.)   22923-8

AN ATLAS OF ANATOMY FOR ARTISTS, Fritz Schider. Most thorough reference work on art anatomy in the world. Hundreds of illustrations, including selections from works by Vesalius, Leonardo, Goya, Ingres, Michelangelo, others. 593 illustrations. 192pp. 7⅛ x 10¼.   20241-0

CELTIC HAND STROKE-BY-STROKE (Irish Half-Uncial from "The Book of Kells"): An Arthur Baker Calligraphy Manual, Arthur Baker. Complete guide to creating each letter of the alphabet in distinctive Celtic manner. Covers hand position, strokes, pens, inks, paper, more. Illustrated. 48pp. 8¼ x 11.   24336-2

EASY ORIGAMI, John Montroll. Charming collection of 32 projects (hat, cup, pelican, piano, swan, many more) specially designed for the novice origami hobbyist. Clearly illustrated easy-to-follow instructions insure that even beginning papercrafters will achieve successful results. 48pp. 8¼ x 11.   27298-2

THE COMPLETE BOOK OF BIRDHOUSE CONSTRUCTION FOR WOOD-WORKERS, Scott D. Campbell. Detailed instructions, illustrations, tables. Also data on bird habitat and instinct patterns. Bibliography. 3 tables. 63 illustrations in 15 figures. 48pp. 5¼ x 8½.   24407-5

BLOOMINGDALE'S ILLUSTRATED 1886 CATALOG: Fashions, Dry Goods and Housewares, Bloomingdale Brothers. Famed merchants' extremely rare catalog depicting about 1,700 products: clothing, housewares, firearms, dry goods, jewelry, more. Invaluable for dating, identifying vintage items. Also, copyright-free graphics for artists, designers. Co-published with Henry Ford Museum & Greenfield Village. 160pp. 8¼ x 11.   25780-0

HISTORIC COSTUME IN PICTURES, Braun & Schneider. Over 1,450 costumed figures in clearly detailed engravings–from dawn of civilization to end of 19th century. Captions. Many folk costumes. 256pp. 8⅜ x 11¾.   23150-X

CATALOG OF DOVER BOOKS

STICKLEY CRAFTSMAN FURNITURE CATALOGS, Gustav Stickley and L. & J. G. Stickley. Beautiful, functional furniture in two authentic catalogs from 1910. 594 illustrations, including 277 photos, show settles, rockers, armchairs, reclining chairs, bookcases, desks, tables. 183pp. 6½ x 9¼. 23838-5

AMERICAN LOCOMOTIVES IN HISTORIC PHOTOGRAPHS: 1858 to 1949, Ron Ziel (ed.). A rare collection of 126 meticulously detailed official photographs, called "builder portraits," of American locomotives that majestically chronicle the rise of steam locomotive power in America. Introduction. Detailed captions. xi+ 129pp. 9 x 12. 27393-8

AMERICA'S LIGHTHOUSES: An Illustrated History, Francis Ross Holland, Jr. Delightfully written, profusely illustrated fact-filled survey of over 200 American lighthouses since 1716. History, anecdotes, technological advances, more. 240pp. 8 x 10¾. 25576-X

TOWARDS A NEW ARCHITECTURE, Le Corbusier. Pioneering manifesto by founder of "International School." Technical and aesthetic theories, views of industry, economics, relation of form to function, "mass-production split" and much more. Profusely illustrated. 320pp. 6⅛ x 9¼. (Available in U.S. only.) 25023-7

HOW THE OTHER HALF LIVES, Jacob Riis. Famous journalistic record, exposing poverty and degradation of New York slums around 1900, by major social reformer. 100 striking and influential photographs. 233pp. 10 x 7⅞. 22012-5

FRUIT KEY AND TWIG KEY TO TREES AND SHRUBS, William M. Harlow. One of the handiest and most widely used identification aids. Fruit key covers 120 deciduous and evergreen species; twig key 160 deciduous species. Easily used. Over 300 photographs. 126pp. 5⅜ x 8½. 20511-8

COMMON BIRD SONGS, Dr. Donald J. Borror. Songs of 60 most common U.S. birds: robins, sparrows, cardinals, bluejays, finches, more–arranged in order of increasing complexity. Up to 9 variations of songs of each species.
Cassette and manual 99911-4

ORCHIDS AS HOUSE PLANTS, Rebecca Tyson Northen. Grow cattleyas and many other kinds of orchids–in a window, in a case, or under artificial light. 63 illustrations. 148pp. 5⅜ x 8½. 23261-1

MONSTER MAZES, Dave Phillips. Masterful mazes at four levels of difficulty. Avoid deadly perils and evil creatures to find magical treasures. Solutions for all 32 exciting illustrated puzzles. 48pp. 8¼ x 11. 26005-4

MOZART'S DON GIOVANNI (DOVER OPERA LIBRETTO SERIES), Wolfgang Amadeus Mozart. Introduced and translated by Ellen H. Bleiler. Standard Italian libretto, with complete English translation. Convenient and thoroughly portable–an ideal companion for reading along with a recording or the performance itself. Introduction. List of characters. Plot summary. 121pp. 5¼ x 8½. 24944-1

TECHNICAL MANUAL AND DICTIONARY OF CLASSICAL BALLET, Gail Grant. Defines, explains, comments on steps, movements, poses and concepts. 15-page pictorial section. Basic book for student, viewer. 127pp. 5⅜ x 8½. 21843-0

THE CLARINET AND CLARINET PLAYING, David Pino. Lively, comprehensive work features suggestions about technique, musicianship, and musical interpretation, as well as guidelines for teaching, making your own reeds, and preparing for public performance. Includes an intriguing look at clarinet history. "A godsend," *The Clarinet,* Journal of the International Clarinet Society. Appendixes. 7 illus. 320pp. 5⅜ x 8½. 40270-3

HOLLYWOOD GLAMOR PORTRAITS, John Kobal (ed.). 145 photos from 1926-49. Harlow, Gable, Bogart, Bacall; 94 stars in all. Full background on photographers, technical aspects. 160pp. 8⅜ x 11¼. 23352-9

THE ANNOTATED CASEY AT THE BAT: A Collection of Ballads about the Mighty Casey/Third, Revised Edition, Martin Gardner (ed.). Amusing sequels and parodies of one of America's best-loved poems: Casey's Revenge, Why Casey Whiffed, Casey's Sister at the Bat, others. 256pp. 5⅜ x 8½. 28598-7

THE RAVEN AND OTHER FAVORITE POEMS, Edgar Allan Poe. Over 40 of the author's most memorable poems: "The Bells," "Ulalume," "Israfel," "To Helen," "The Conqueror Worm," "Eldorado," "Annabel Lee," many more. Alphabetic lists of titles and first lines. 64pp. 5$\frac{3}{16}$ x 8¼. 26685-0

PERSONAL MEMOIRS OF U. S. GRANT, Ulysses Simpson Grant. Intelligent, deeply moving firsthand account of Civil War campaigns, considered by many the finest military memoirs ever written. Includes letters, historic photographs, maps and more. 528pp. 6⅛ x 9¼. 28587-1

ANCIENT EGYPTIAN MATERIALS AND INDUSTRIES, A. Lucas and J. Harris. Fascinating, comprehensive, thoroughly documented text describes this ancient civilization's vast resources and the processes that incorporated them in daily life, including the use of animal products, building materials, cosmetics, perfumes and incense, fibers, glazed ware, glass and its manufacture, materials used in the mummification process, and much more. 544pp. 6⅛ x 9¼. (Available in U.S. only.) 40446-3

RUSSIAN STORIES/RUSSKIE RASSKAZY: A Dual-Language Book, edited by Gleb Struve. Twelve tales by such masters as Chekhov, Tolstoy, Dostoevsky, Pushkin, others. Excellent word-for-word English translations on facing pages, plus teaching and study aids, Russian/English vocabulary, biographical/critical introductions, more. 416pp. 5⅜ x 8½. 26244-8

PHILADELPHIA THEN AND NOW: 60 Sites Photographed in the Past and Present, Kenneth Finkel and Susan Oyama. Rare photographs of City Hall, Logan Square, Independence Hall, Betsy Ross House, other landmarks juxtaposed with contemporary views. Captures changing face of historic city. Introduction. Captions. 128pp. 8¼ x 11. 25790-8

AIA ARCHITECTURAL GUIDE TO NASSAU AND SUFFOLK COUNTIES, LONG ISLAND, The American Institute of Architects, Long Island Chapter, and the Society for the Preservation of Long Island Antiquities. Comprehensive, well-researched and generously illustrated volume brings to life over three centuries of Long Island's great architectural heritage. More than 240 photographs with authoritative, extensively detailed captions. 176pp. 8¼ x 11. 26946-9

NORTH AMERICAN INDIAN LIFE: Customs and Traditions of 23 Tribes, Elsie Clews Parsons (ed.). 27 fictionalized essays by noted anthropologists examine religion, customs, government, additional facets of life among the Winnebago, Crow, Zuni, Eskimo, other tribes. 480pp. 6⅛ x 9¼. 27377-6

FRANK LLOYD WRIGHT'S DANA HOUSE, Donald Hoffmann. Pictorial essay of residential masterpiece with over 160 interior and exterior photos, plans, elevations, sketches and studies. 128pp. 9¼ x 10¾. 29120-0

THE MALE AND FEMALE FIGURE IN MOTION: 60 Classic Photographic Sequences, Eadweard Muybridge. 60 true-action photographs of men and women walking, running, climbing, bending, turning, etc., reproduced from rare 19th-century masterpiece. vi + 121pp. 9 x 12. 24745-7

1001 QUESTIONS ANSWERED ABOUT THE SEASHORE, N. J. Berrill and Jacquelyn Berrill. Queries answered about dolphins, sea snails, sponges, starfish, fishes, shore birds, many others. Covers appearance, breeding, growth, feeding, much more. 305pp. 5¼ x 8¼. 23366-9

ATTRACTING BIRDS TO YOUR YARD, William J. Weber. Easy-to-follow guide offers advice on how to attract the greatest diversity of birds: birdhouses, feeders, water and waterers, much more. 96pp. 5³⁄₁₆ x 8¼. 28927-3

MEDICINAL AND OTHER USES OF NORTH AMERICAN PLANTS: A Historical Survey with Special Reference to the Eastern Indian Tribes, Charlotte Erichsen-Brown. Chronological historical citations document 500 years of usage of plants, trees, shrubs native to eastern Canada, northeastern U.S. Also complete identifying information. 343 illustrations. 544pp. 6½ x 9¼. 25951-X

STORYBOOK MAZES, Dave Phillips. 23 stories and mazes on two-page spreads: Wizard of Oz, Treasure Island, Robin Hood, etc. Solutions. 64pp. 8¼ x 11. 23628-5

AMERICAN NEGRO SONGS: 230 Folk Songs and Spirituals, Religious and Secular, John W. Work. This authoritative study traces the African influences of songs sung and played by black Americans at work, in church, and as entertainment. The author discusses the lyric significance of such songs as "Swing Low, Sweet Chariot," "John Henry," and others and offers the words and music for 230 songs. Bibliography. Index of Song Titles. 272pp. 6½ x 9¼. 40271-1

MOVIE-STAR PORTRAITS OF THE FORTIES, John Kobal (ed.). 163 glamor, studio photos of 106 stars of the 1940s: Rita Hayworth, Ava Gardner, Marlon Brando, Clark Gable, many more. 176pp. 8⅜ x 11¼. 23546-7

BENCHLEY LOST AND FOUND, Robert Benchley. Finest humor from early 30s, about pet peeves, child psychologists, post office and others. Mostly unavailable elsewhere. 73 illustrations by Peter Arno and others. 183pp. 5⅜ x 8½. 22410-4

YEKL and THE IMPORTED BRIDEGROOM AND OTHER STORIES OF YIDDISH NEW YORK, Abraham Cahan. Film Hester Street based on *Yekl* (1896). Novel, other stories among first about Jewish immigrants on N.Y.'s East Side. 240pp. 5⅜ x 8½. 22427-9

SELECTED POEMS, Walt Whitman. Generous sampling from *Leaves of Grass*. Twenty-four poems include "I Hear America Singing," "Song of the Open Road," "I Sing the Body Electric," "When Lilacs Last in the Dooryard Bloom'd," "O Captain! My Captain!"—all reprinted from an authoritative edition. Lists of titles and first lines. 128pp. 5³⁄₁₆ x 8¼. 26878-0

THE BEST TALES OF HOFFMANN, E. T. A. Hoffmann. 10 of Hoffmann's most important stories: "Nutcracker and the King of Mice," "The Golden Flowerpot," etc. 458pp. 5⅜ x 8½. 21793-0

FROM FETISH TO GOD IN ANCIENT EGYPT, E. A. Wallis Budge. Rich detailed survey of Egyptian conception of "God" and gods, magic, cult of animals, Osiris, more. Also, superb English translations of hymns and legends. 240 illustrations. 545pp. 5⅜ x 8½. 25803-3

FRENCH STORIES/CONTES FRANÇAIS: A Dual-Language Book, Wallace Fowlie. Ten stories by French masters, Voltaire to Camus: "Micromegas" by Voltaire; "The Atheist's Mass" by Balzac; "Minuet" by de Maupassant; "The Guest" by Camus, six more. Excellent English translations on facing pages. Also French-English vocabulary list, exercises, more. 352pp. 5⅜ x 8½. 26443-2

CHICAGO AT THE TURN OF THE CENTURY IN PHOTOGRAPHS: 122 Historic Views from the Collections of the Chicago Historical Society, Larry A. Viskochil. Rare large-format prints offer detailed views of City Hall, State Street, the Loop, Hull House, Union Station, many other landmarks, circa 1904-1913. Introduction. Captions. Maps. 144pp. 9⅜ x 12¼. 24656-6

OLD BROOKLYN IN EARLY PHOTOGRAPHS, 1865-1929, William Lee Younger. Luna Park, Gravesend race track, construction of Grand Army Plaza, moving of Hotel Brighton, etc. 157 previously unpublished photographs. 165pp. 8⅞ x 11¾. 23587-4

THE MYTHS OF THE NORTH AMERICAN INDIANS, Lewis Spence. Rich anthology of the myths and legends of the Algonquins, Iroquois, Pawnees and Sioux, prefaced by an extensive historical and ethnological commentary. 36 illustrations. 480pp. 5⅜ x 8½. 25967-6

AN ENCYCLOPEDIA OF BATTLES: Accounts of Over 1,560 Battles from 1479 B.C. to the Present, David Eggenberger. Essential details of every major battle in recorded history from the first battle of Megiddo in 1479 B.C. to Grenada in 1984. List of Battle Maps. New Appendix covering the years 1967-1984. Index. 99 illustrations. 544pp. 6½ x 9¼. 24913-1

SAILING ALONE AROUND THE WORLD, Captain Joshua Slocum. First man to sail around the world, alone, in small boat. One of great feats of seamanship told in delightful manner. 67 illustrations. 294pp. 5⅜ x 8½. 20326-3

ANARCHISM AND OTHER ESSAYS, Emma Goldman. Powerful, penetrating, prophetic essays on direct action, role of minorities, prison reform, puritan hypocrisy, violence, etc. 271pp. 5⅜ x 8½. 22484-8

MYTHS OF THE HINDUS AND BUDDHISTS, Ananda K. Coomaraswamy and Sister Nivedita. Great stories of the epics; deeds of Krishna, Shiva, taken from puranas, Vedas, folk tales; etc. 32 illustrations. 400pp. 5⅜ x 8½. 21759-0

THE TRAUMA OF BIRTH, Otto Rank. Rank's controversial thesis that anxiety neurosis is caused by profound psychological trauma which occurs at birth. 256pp. 5⅜ x 8½. 27974-X

A THEOLOGICO-POLITICAL TREATISE, Benedict Spinoza. Also contains unfinished Political Treatise. Great classic on religious liberty, theory of government on common consent. R. Elwes translation. Total of 421pp. 5⅜ x 8½. 20249-6

CATALOG OF DOVER BOOKS

MY BONDAGE AND MY FREEDOM, Frederick Douglass. Born a slave, Douglass became outspoken force in antislavery movement. The best of Douglass' autobiographies. Graphic description of slave life. 464pp. 5⅜ x 8½. 22457-0

FOLLOWING THE EQUATOR: A Journey Around the World, Mark Twain. Fascinating humorous account of 1897 voyage to Hawaii, Australia, India, New Zealand, etc. Ironic, bemused reports on peoples, customs, climate, flora and fauna, politics, much more. 197 illustrations. 720pp. 5⅜ x 8½. 26113-1

THE PEOPLE CALLED SHAKERS, Edward D. Andrews. Definitive study of Shakers: origins, beliefs, practices, dances, social organization, furniture and crafts, etc. 33 illustrations. 351pp. 5⅜ x 8½. 21081-2

THE MYTHS OF GREECE AND ROME, H. A. Guerber. A classic of mythology, generously illustrated, long prized for its simple, graphic, accurate retelling of the principal myths of Greece and Rome, and for its commentary on their origins and significance. With 64 illustrations by Michelangelo, Raphael, Titian, Rubens, Canova, Bernini and others. 480pp. 5⅜ x 8½. 27584-1

PSYCHOLOGY OF MUSIC, Carl E. Seashore. Classic work discusses music as a medium from psychological viewpoint. Clear treatment of physical acoustics, auditory apparatus, sound perception, development of musical skills, nature of musical feeling, host of other topics. 88 figures. 408pp. 5⅜ x 8½. 21851-1

THE PHILOSOPHY OF HISTORY, Georg W. Hegel. Great classic of Western thought develops concept that history is not chance but rational process, the evolution of freedom. 457pp. 5⅜ x 8½. 20112-0

THE BOOK OF TEA, Kakuzo Okakura. Minor classic of the Orient: entertaining, charming explanation, interpretation of traditional Japanese culture in terms of tea ceremony. 94pp. 5⅜ x 8½. 20070-1

LIFE IN ANCIENT EGYPT, Adolf Erman. Fullest, most thorough, detailed older account with much not in more recent books, domestic life, religion, magic, medicine, commerce, much more. Many illustrations reproduce tomb paintings, carvings, hieroglyphs, etc. 597pp. 5⅜ x 8½. 22632-8

SUNDIALS, Their Theory and Construction, Albert Waugh. Far and away the best, most thorough coverage of ideas, mathematics concerned, types, construction, adjusting anywhere. Simple, nontechnical treatment allows even children to build several of these dials. Over 100 illustrations. 230pp. 5⅜ x 8½. 22947-5

THEORETICAL HYDRODYNAMICS, L. M. Milne-Thomson. Classic exposition of the mathematical theory of fluid motion, applicable to both hydrodynamics and aerodynamics. Over 600 exercises. 768pp. 6⅛ x 9¼. 68970-0

SONGS OF EXPERIENCE: Facsimile Reproduction with 26 Plates in Full Color, William Blake. 26 full-color plates from a rare 1826 edition. Includes "The Tyger," "London," "Holy Thursday," and other poems. Printed text of poems. 48pp. 5¼ x 7. 24636-1

OLD-TIME VIGNETTES IN FULL COLOR, Carol Belanger Grafton (ed.). Over 390 charming, often sentimental illustrations, selected from archives of Victorian graphics—pretty women posing, children playing, food, flowers, kittens and puppies, smiling cherubs, birds and butterflies, much more. All copyright-free. 48pp. 9¼ x 12¼. 27269-9

PERSPECTIVE FOR ARTISTS, Rex Vicat Cole. Depth, perspective of sky and sea, shadows, much more, not usually covered. 391 diagrams, 81 reproductions of drawings and paintings. 279pp. 5⅜ x 8½. 22487-2

DRAWING THE LIVING FIGURE, Joseph Sheppard. Innovative approach to artistic anatomy focuses on specifics of surface anatomy, rather than muscles and bones. Over 170 drawings of live models in front, back and side views, and in widely varying poses. Accompanying diagrams. 177 illustrations. Introduction. Index. 144pp. 8⅜ x11¼. 26723-7

GOTHIC AND OLD ENGLISH ALPHABETS: 100 Complete Fonts, Dan X. Solo. Add power, elegance to posters, signs, other graphics with 100 stunning copyright-free alphabets: Blackstone, Dolbey, Germania, 97 more—including many lower-case, numerals, punctuation marks. 104pp. 8⅛ x 11. 24695-7

HOW TO DO BEADWORK, Mary White. Fundamental book on craft from simple projects to five-bead chains and woven works. 106 illustrations. 142pp. 5⅜ x 8.

20697-1

THE BOOK OF WOOD CARVING, Charles Marshall Sayers. Finest book for beginners discusses fundamentals and offers 34 designs. "Absolutely first rate . . . well thought out and well executed."—E. J. Tangerman. 118pp. 7¾ x 10⅝. 23654-4

ILLUSTRATED CATALOG OF CIVIL WAR MILITARY GOODS: Union Army Weapons, Insignia, Uniform Accessories, and Other Equipment, Schuyler, Hartley, and Graham. Rare, profusely illustrated 1846 catalog includes Union Army uniform and dress regulations, arms and ammunition, coats, insignia, flags, swords, rifles, etc. 226 illustrations. 160pp. 9 x 12. 24939-5

WOMEN'S FASHIONS OF THE EARLY 1900s: An Unabridged Republication of "New York Fashions, 1909," National Cloak & Suit Co. Rare catalog of mail-order fashions documents women's and children's clothing styles shortly after the turn of the century. Captions offer full descriptions, prices. Invaluable resource for fashion, costume historians. Approximately 725 illustrations. 128pp. 8⅜ x 11¼. 27276-1

THE 1912 AND 1915 GUSTAV STICKLEY FURNITURE CATALOGS, Gustav Stickley. With over 200 detailed illustrations and descriptions, these two catalogs are essential reading and reference materials and identification guides for Stickley furniture. Captions cite materials, dimensions and prices. 112pp. 6½ x 9¼. 26676-1

EARLY AMERICAN LOCOMOTIVES, John H. White, Jr. Finest locomotive engravings from early 19th century: historical (1804–74), main-line (after 1870), special, foreign, etc. 147 plates. 142pp. 11⅜ x 8¼. 22772-3

THE TALL SHIPS OF TODAY IN PHOTOGRAPHS, Frank O. Braynard. Lavishly illustrated tribute to nearly 100 majestic contemporary sailing vessels: Amerigo Vespucci, Clearwater, Constitution, Eagle, Mayflower, Sea Cloud, Victory, many more. Authoritative captions provide statistics, background on each ship. 190 black-and-white photographs and illustrations. Introduction. 128pp. 8⅞ x 11¾. 27163-3

CATALOG OF DOVER BOOKS

LITTLE BOOK OF EARLY AMERICAN CRAFTS AND TRADES, Peter Stockham (ed.). 1807 children's book explains crafts and trades: baker, hatter, cooper, potter, and many others. 23 copperplate illustrations. 140pp. 4⅝ x 6.   23336-7

VICTORIAN FASHIONS AND COSTUMES FROM HARPER'S BAZAR, 1867–1898, Stella Blum (ed.). Day costumes, evening wear, sports clothes, shoes, hats, other accessories in over 1,000 detailed engravings. 320pp. 9⅜ x 12¼. 22990-4

GUSTAV STICKLEY, THE CRAFTSMAN, Mary Ann Smith. Superb study surveys broad scope of Stickley's achievement, especially in architecture. Design philosophy, rise and fall of the Craftsman empire, descriptions and floor plans for many Craftsman houses, more. 86 black-and-white halftones. 31 line illustrations. Introduction 208pp. 6½ x 9¼.   27210-9

THE LONG ISLAND RAIL ROAD IN EARLY PHOTOGRAPHS, Ron Ziel. Over 220 rare photos, informative text document origin (1844) and development of rail service on Long Island. Vintage views of early trains, locomotives, stations, passengers, crews, much more. Captions. 8⅞ x 11¾.   26301-0

VOYAGE OF THE LIBERDADE, Joshua Slocum. Great 19th-century mariner's thrilling, first-hand account of the wreck of his ship off South America, the 35-foot boat he built from the wreckage, and its remarkable voyage home. 128pp. 5⅜ x 8½.
40022-0

TEN BOOKS ON ARCHITECTURE, Vitruvius. The most important book ever written on architecture. Early Roman aesthetics, technology, classical orders, site selection, all other aspects. Morgan translation. 331pp. 5⅜ x 8½.   20645-9

THE HUMAN FIGURE IN MOTION, Eadweard Muybridge. More than 4,500 stopped-action photos, in action series, showing undraped men, women, children jumping, lying down, throwing, sitting, wrestling, carrying, etc. 390pp. 7⅞ x 10⅝.
20204-6 Clothbd.

TREES OF THE EASTERN AND CENTRAL UNITED STATES AND CANADA, William M. Harlow. Best one-volume guide to 140 trees. Full descriptions, woodlore, range, etc. Over 600 illustrations. Handy size. 288pp. 4½ x 6⅜.   20395-6

SONGS OF WESTERN BIRDS, Dr. Donald J. Borror. Complete song and call repertoire of 60 western species, including flycatchers, juncoes, cactus wrens, many more—includes fully illustrated booklet.   Cassette and manual 99913-0

GROWING AND USING HERBS AND SPICES, Milo Miloradovich. Versatile handbook provides all the information needed for cultivation and use of all the herbs and spices available in North America. 4 illustrations. Index. Glossary. 236pp. 5⅜ x 8½.
25058-X

BIG BOOK OF MAZES AND LABYRINTHS, Walter Shepherd. 50 mazes and labyrinths in all—classical, solid, ripple, and more—in one great volume. Perfect inexpensive puzzler for clever youngsters. Full solutions. 112pp. 8⅛ x 11.   22951-3

PIANO TUNING, J. Cree Fischer. Clearest, best book for beginner, amateur. Simple repairs, raising dropped notes, tuning by easy method of flattened fifths. No previous skills needed. 4 illustrations. 201pp. 5⅜ x 8½. 23267-0

HINTS TO SINGERS, Lillian Nordica. Selecting the right teacher, developing confidence, overcoming stage fright, and many other important skills receive thoughtful discussion in this indispensible guide, written by a world-famous diva of four decades' experience. 96pp. 5⅜ x 8½. 40094-8

THE COMPLETE NONSENSE OF EDWARD LEAR, Edward Lear. All nonsense limericks, zany alphabets, Owl and Pussycat, songs, nonsense botany, etc., illustrated by Lear. Total of 320pp. 5⅜ x 8½. (Available in U.S. only.) 20167-8

VICTORIAN PARLOUR POETRY: An Annotated Anthology, Michael R. Turner. 117 gems by Longfellow, Tennyson, Browning, many lesser-known poets. "The Village Blacksmith," "Curfew Must Not Ring Tonight," "Only a Baby Small," dozens more, often difficult to find elsewhere. Index of poets, titles, first lines. xxiii + 325pp. 5⅝ x 8¼. 27044-0

DUBLINERS, James Joyce. Fifteen stories offer vivid, tightly focused observations of the lives of Dublin's poorer classes. At least one, "The Dead," is considered a masterpiece. Reprinted complete and unabridged from standard edition. 160pp. 5³⁄₁₆ x 8¼. 26870-5

GREAT WEIRD TALES: 14 Stories by Lovecraft, Blackwood, Machen and Others, S. T. Joshi (ed.). 14 spellbinding tales, including "The Sin Eater," by Fiona McLeod, "The Eye Above the Mantel," by Frank Belknap Long, as well as renowned works by R. H. Barlow, Lord Dunsany, Arthur Machen, W. C. Morrow and eight other masters of the genre. 256pp. 5⅜ x 8½. (Available in U.S. only.) 40436-6

THE BOOK OF THE SACRED MAGIC OF ABRAMELIN THE MAGE, translated by S. MacGregor Mathers. Medieval manuscript of ceremonial magic. Basic document in Aleister Crowley, Golden Dawn groups. 268pp. 5⅜ x 8½. 23211-5

NEW RUSSIAN-ENGLISH AND ENGLISH-RUSSIAN DICTIONARY, M. A. O'Brien. This is a remarkably handy Russian dictionary, containing a surprising amount of information, including over 70,000 entries. 366pp. 4½ x 6⅛. 20208-9

HISTORIC HOMES OF THE AMERICAN PRESIDENTS, Second, Revised Edition, Irvin Haas. A traveler's guide to American Presidential homes, most open to the public, depicting and describing homes occupied by every American President from George Washington to George Bush. With visiting hours, admission charges, travel routes. 175 photographs. Index. 160pp. 8¼ x 11. 26751-2

NEW YORK IN THE FORTIES, Andreas Feininger. 162 brilliant photographs by the well-known photographer, formerly with *Life* magazine. Commuters, shoppers, Times Square at night, much else from city at its peak. Captions by John von Hartz. 181pp. 9¼ x 10¾. 23585-8

INDIAN SIGN LANGUAGE, William Tomkins. Over 525 signs developed by Sioux and other tribes. Written instructions and diagrams. Also 290 pictographs. 111pp. 6⅛ x 9¼. 22029-X

CATALOG OF DOVER BOOKS

ANATOMY: A Complete Guide for Artists, Joseph Sheppard. A master of figure drawing shows artists how to render human anatomy convincingly. Over 460 illustrations. 224pp. 8⅜ x 11¼. 27279-6

MEDIEVAL CALLIGRAPHY: Its History and Technique, Marc Drogin. Spirited history, comprehensive instruction manual covers 13 styles (ca. 4th century through 15th). Excellent photographs; directions for duplicating medieval techniques with modern tools. 224pp. 8⅜ x 11¼. 26142-5

DRIED FLOWERS: How to Prepare Them, Sarah Whitlock and Martha Rankin. Complete instructions on how to use silica gel, meal and borax, perlite aggregate, sand and borax, glycerine and water to create attractive permanent flower arrangements. 12 illustrations. 32pp. 5⅜ x 8½. 21802-3

EASY-TO-MAKE BIRD FEEDERS FOR WOODWORKERS, Scott D. Campbell. Detailed, simple-to-use guide for designing, constructing, caring for and using feeders. Text, illustrations for 12 classic and contemporary designs. 96pp. 5⅜ x 8½. 25847-5

SCOTTISH WONDER TALES FROM MYTH AND LEGEND, Donald A. Mackenzie. 16 lively tales tell of giants rumbling down mountainsides, of a magic wand that turns stone pillars into warriors, of gods and goddesses, evil hags, powerful forces and more. 240pp. 5⅜ x 8½. 29677-6

THE HISTORY OF UNDERCLOTHES, C. Willett Cunnington and Phyllis Cunnington. Fascinating, well-documented survey covering six centuries of English undergarments, enhanced with over 100 illustrations: 12th-century laced-up bodice, footed long drawers (1795), 19th-century bustles, l9th-century corsets for men, Victorian "bust improvers," much more. 272pp. 5⅜ x 8¼. 27124-2

ARTS AND CRAFTS FURNITURE: The Complete Brooks Catalog of 1912, Brooks Manufacturing Co. Photos and detailed descriptions of more than 150 now very collectible furniture designs from the Arts and Crafts movement depict davenports, settees, buffets, desks, tables, chairs, bedsteads, dressers and more, all built of solid, quarter-sawed oak. Invaluable for students and enthusiasts of antiques, Americana and the decorative arts. 80pp. 6½ x 9¼. 27471-3

WILBUR AND ORVILLE: A Biography of the Wright Brothers, Fred Howard. Definitive, crisply written study tells the full story of the brothers' lives and work. A vividly written biography, unparalleled in scope and color, that also captures the spirit of an extraordinary era. 560pp. 6⅛ x 9¼. 40297-5

THE ARTS OF THE SAILOR: Knotting, Splicing and Ropework, Hervey Garrett Smith. Indispensable shipboard reference covers tools, basic knots and useful hitches; handsewing and canvas work, more. Over 100 illustrations. Delightful reading for sea lovers. 256pp. 5⅜ x 8½. 26440-8

FRANK LLOYD WRIGHT'S FALLINGWATER: The House and Its History, Second, Revised Edition, Donald Hoffmann. A total revision—both in text and illustrations—of the standard document on Fallingwater, the boldest, most personal architectural statement of Wright's mature years, updated with valuable new material from the recently opened Frank Lloyd Wright Archives. "Fascinating"—*The New York Times*. 116 illustrations. 128pp. 9¼ x 10¾. 27430-6

CATALOG OF DOVER BOOKS

PHOTOGRAPHIC SKETCHBOOK OF THE CIVIL WAR, Alexander Gardner. 100 photos taken on field during the Civil War. Famous shots of Manassas Harper's Ferry, Lincoln, Richmond, slave pens, etc. 244pp. 10⅞ x 8¼. 22731-6

FIVE ACRES AND INDEPENDENCE, Maurice G. Kains. Great back-to-the-land classic explains basics of self-sufficient farming. The one book to get. 95 illustrations. 397pp. 5⅜ x 8½. 20974-1

SONGS OF EASTERN BIRDS, Dr. Donald J. Borror. Songs and calls of 60 species most common to eastern U.S.: warblers, woodpeckers, flycatchers, thrushes, larks, many more in high-quality recording. Cassette and manual 99912-2

A MODERN HERBAL, Margaret Grieve. Much the fullest, most exact, most useful compilation of herbal material. Gigantic alphabetical encyclopedia, from aconite to zedoary, gives botanical information, medical properties, folklore, economic uses, much else. Indispensable to serious reader. 161 illustrations. 888pp. 6½ x 9¼. 2-vol. set. (Available in U.S. only.) Vol. I: 22798-7
Vol. II: 22799-5

HIDDEN TREASURE MAZE BOOK, Dave Phillips. Solve 34 challenging mazes accompanied by heroic tales of adventure. Evil dragons, people-eating plants, bloodthirsty giants, many more dangerous adversaries lurk at every twist and turn. 34 mazes, stories, solutions. 48pp. 8¼ x 11. 24566-7

LETTERS OF W. A. MOZART, Wolfgang A. Mozart. Remarkable letters show bawdy wit, humor, imagination, musical insights, contemporary musical world; includes some letters from Leopold Mozart. 276pp. 5⅜ x 8½. 22859-2

BASIC PRINCIPLES OF CLASSICAL BALLET, Agrippina Vaganova. Great Russian theoretician, teacher explains methods for teaching classical ballet. 118 illustrations. 175pp. 5⅜ x 8½. 22036-2

THE JUMPING FROG, Mark Twain. Revenge edition. The original story of The Celebrated Jumping Frog of Calaveras County, a hapless French translation, and Twain's hilarious "retranslation" from the French. 12 illustrations. 66pp. 5⅜ x 8½. 22686-7

BEST REMEMBERED POEMS, Martin Gardner (ed.). The 126 poems in this superb collection of 19th- and 20th-century British and American verse range from Shelley's "To a Skylark" to the impassioned "Renascence" of Edna St. Vincent Millay and to Edward Lear's whimsical "The Owl and the Pussycat." 224pp. 5⅜ x 8½. 27165-X

COMPLETE SONNETS, William Shakespeare. Over 150 exquisite poems deal with love, friendship, the tyranny of time, beauty's evanescence, death and other themes in language of remarkable power, precision and beauty. Glossary of archaic terms. 80pp. 5³⁄₁₆ x 8¼. 26686-9

THE BATTLES THAT CHANGED HISTORY, Fletcher Pratt. Eminent historian profiles 16 crucial conflicts, ancient to modern, that changed the course of civilization. 352pp. 5⅜ x 8½. 41129-X

THE WIT AND HUMOR OF OSCAR WILDE, Alvin Redman (ed.). More than 1,000 ripostes, paradoxes, wisecracks: Work is the curse of the drinking classes; I can resist everything except temptation; etc. 258pp. 5⅜ x 8½. 20602-5

SHAKESPEARE LEXICON AND QUOTATION DICTIONARY, Alexander Schmidt. Full definitions, locations, shades of meaning in every word in plays and poems. More than 50,000 exact quotations. 1,485pp. 6½ x 9¼. 2-vol. set.
Vol. 1: 22726-X
Vol. 2: 22727-8

SELECTED POEMS, Emily Dickinson. Over 100 best-known, best-loved poems by one of America's foremost poets, reprinted from authoritative early editions. No comparable edition at this price. Index of first lines. 64pp. 5$\frac{3}{16}$ x 8¼. 26466-1

THE INSIDIOUS DR. FU-MANCHU, Sax Rohmer. The first of the popular mystery series introduces a pair of English detectives to their archnemesis, the diabolical Dr. Fu-Manchu. Flavorful atmosphere, fast-paced action, and colorful characters enliven this classic of the genre. 208pp. 5$\frac{3}{16}$ x 8¼. 29898-1

THE MALLEUS MALEFICARUM OF KRAMER AND SPRENGER, translated by Montague Summers. Full text of most important witchhunter's "bible," used by both Catholics and Protestants. 278pp. 6⅝ x 10. 22802-9

SPANISH STORIES/CUENTOS ESPAÑOLES: A Dual-Language Book, Angel Flores (ed.). Unique format offers 13 great stories in Spanish by Cervantes, Borges, others. Faithful English translations on facing pages. 352pp. 5⅜ x 8½. 25399-6

GARDEN CITY, LONG ISLAND, IN EARLY PHOTOGRAPHS, 1869–1919, Mildred H. Smith. Handsome treasury of 118 vintage pictures, accompanied by carefully researched captions, document the Garden City Hotel fire (1899), the Vanderbilt Cup Race (1908), the first airmail flight departing from the Nassau Boulevard Aerodrome (1911), and much more. 96pp. 8⅞ x 11¾. 40669-5

OLD QUEENS, N.Y., IN EARLY PHOTOGRAPHS, Vincent F. Seyfried and William Asadorian. Over 160 rare photographs of Maspeth, Jamaica, Jackson Heights, and other areas. Vintage views of DeWitt Clinton mansion, 1939 World's Fair and more. Captions. 192pp. 8⅞ x 11. 26358-4

CAPTURED BY THE INDIANS: 15 Firsthand Accounts, 1750-1870, Frederick Drimmer. Astounding true historical accounts of grisly torture, bloody conflicts, relentless pursuits, miraculous escapes and more, by people who lived to tell the tale. 384pp. 5⅜ x 8½. 24901-8

THE WORLD'S GREAT SPEECHES (Fourth Enlarged Edition), Lewis Copeland, Lawrence W. Lamm, and Stephen J. McKenna. Nearly 300 speeches provide public speakers with a wealth of updated quotes and inspiration–from Pericles' funeral oration and William Jennings Bryan's "Cross of Gold Speech" to Malcolm X's powerful words on the Black Revolution and Earl of Spenser's tribute to his sister, Diana, Princess of Wales. 944pp. 5⅜ x 8⅜. 40903-1

THE BOOK OF THE SWORD, Sir Richard F. Burton. Great Victorian scholar/adventurer's eloquent, erudite history of the "queen of weapons"–from prehistory to early Roman Empire. Evolution and development of early swords, variations (sabre, broadsword, cutlass, scimitar, etc.), much more. 336pp. 6⅛ x 9¼.
25434-8

CATALOG OF DOVER BOOKS

AUTOBIOGRAPHY: The Story of My Experiments with Truth, Mohandas K. Gandhi. Boyhood, legal studies, purification, the growth of the Satyagraha (nonviolent protest) movement. Critical, inspiring work of the man responsible for the freedom of India. 480pp. 5⅜ x 8½. (Available in U.S. only.) 24593-4

CELTIC MYTHS AND LEGENDS, T. W. Rolleston. Masterful retelling of Irish and Welsh stories and tales. Cuchulain, King Arthur, Deirdre, the Grail, many more. First paperback edition. 58 full-page illustrations. 512pp. 5⅜ x 8½. 26507-2

THE PRINCIPLES OF PSYCHOLOGY, William James. Famous long course complete, unabridged. Stream of thought, time perception, memory, experimental methods; great work decades ahead of its time. 94 figures. 1,391pp. 5⅜ x 8½. 2-vol. set.
Vol. I: 20381-6   Vol. II: 20382-4

THE WORLD AS WILL AND REPRESENTATION, Arthur Schopenhauer. Definitive English translation of Schopenhauer's life work, correcting more than 1,000 errors, omissions in earlier translations. Translated by E. F. J. Payne. Total of 1,269pp. 5⅜ x 8½. 2-vol. set. Vol. 1: 21761-2   Vol. 2: 21762-0

MAGIC AND MYSTERY IN TIBET, Madame Alexandra David-Neel. Experiences among lamas, magicians, sages, sorcerers, Bonpa wizards. A true psychic discovery. 32 illustrations. 321pp. 5⅜ x 8½. (Available in U.S. only.) 22682-4

THE EGYPTIAN BOOK OF THE DEAD, E. A. Wallis Budge. Complete reproduction of Ani's papyrus, finest ever found. Full hieroglyphic text, interlinear transliteration, word-for-word translation, smooth translation. 533pp. 6½ x 9¼. 21866-X

MATHEMATICS FOR THE NONMATHEMATICIAN, Morris Kline. Detailed, college-level treatment of mathematics in cultural and historical context, with numerous exercises. Recommended Reading Lists. Tables. Numerous figures. 641pp. 5⅜ x 8½. 24823-2

PROBABILISTIC METHODS IN THE THEORY OF STRUCTURES, Isaac Elishakoff. Well-written introduction covers the elements of the theory of probability from two or more random variables, the reliability of such multivariable structures, the theory of random function, Monte Carlo methods of treating problems incapable of exact solution, and more. Examples. 502pp. 5⅜ x 8½. 40691-1

THE RIME OF THE ANCIENT MARINER, Gustave Doré, S. T. Coleridge. Doré's finest work; 34 plates capture moods, subtleties of poem. Flawless full-size reproductions printed on facing pages with authoritative text of poem. "Beautiful. Simply beautiful."—*Publisher's Weekly.* 77pp. 9¼ x 12. 22305-1

NORTH AMERICAN INDIAN DESIGNS FOR ARTISTS AND CRAFTSPEOPLE, Eva Wilson. Over 360 authentic copyright-free designs adapted from Navajo blankets, Hopi pottery, Sioux buffalo hides, more. Geometrics, symbolic figures, plant and animal motifs, etc. 128pp. 8⅜ x 11. (Not for sale in the United Kingdom.) 25341-4

SCULPTURE: Principles and Practice, Louis Slobodkin. Step-by-step approach to clay, plaster, metals, stone; classical and modern. 253 drawings, photos. 255pp. 8⅛ x 11. 22960-2

THE INFLUENCE OF SEA POWER UPON HISTORY, 1660–1783, A. T. Mahan. Influential classic of naval history and tactics still used as text in war colleges. First paperback edition. 4 maps. 24 battle plans. 640pp. 5⅜ x 8½. 25509-3

CATALOG OF DOVER BOOKS

THE STORY OF THE TITANIC AS TOLD BY ITS SURVIVORS, Jack Winocour (ed.). What it was really like. Panic, despair, shocking inefficiency, and a little heroism. More thrilling than any fictional account. 26 illustrations. 320pp. 5⅜ x 8½.
20610-6

FAIRY AND FOLK TALES OF THE IRISH PEASANTRY, William Butler Yeats (ed.). Treasury of 64 tales from the twilight world of Celtic myth and legend: "The Soul Cages," "The Kildare Pooka," "King O'Toole and his Goose," many more. Introduction and Notes by W. B. Yeats. 352pp. 5⅜ x 8½.
26941-8

BUDDHIST MAHAYANA TEXTS, E. B. Cowell and others (eds.). Superb, accurate translations of basic documents in Mahayana Buddhism, highly important in history of religions. The Buddha-karita of Asvaghosha, Larger Sukhavativyuha, more. 448pp. 5⅜ x 8½.
25552-2

ONE TWO THREE . . . INFINITY: Facts and Speculations of Science, George Gamow. Great physicist's fascinating, readable overview of contemporary science: number theory, relativity, fourth dimension, entropy, genes, atomic structure, much more. 128 illustrations. Index. 352pp. 5⅜ x 8½.
25664-2

EXPERIMENTATION AND MEASUREMENT, W. J. Youden. Introductory manual explains laws of measurement in simple terms and offers tips for achieving accuracy and minimizing errors. Mathematics of measurement, use of instruments, experimenting with machines. 1994 edition. Foreword. Preface. Introduction. Epilogue. Selected Readings. Glossary. Index. Tables and figures. 128pp. 5⅜ x 8½.
40451-X

DALÍ ON MODERN ART: The Cuckolds of Antiquated Modern Art, Salvador Dalí. Influential painter skewers modern art and its practitioners. Outrageous evaluations of Picasso, Cézanne, Turner, more. 15 renderings of paintings discussed. 44 calligraphic decorations by Dalí. 96pp. 5⅜ x 8½. (Available in U.S. only.)
29220-7

ANTIQUE PLAYING CARDS: A Pictorial History, Henry René D'Allemagne. Over 900 elaborate, decorative images from rare playing cards (14th–20th centuries): Bacchus, death, dancing dogs, hunting scenes, royal coats of arms, players cheating, much more. 96pp. 9¼ x 12¼.
29265-7

MAKING FURNITURE MASTERPIECES: 30 Projects with Measured Drawings, Franklin H. Gottshall. Step-by-step instructions, illustrations for constructing handsome, useful pieces, among them a Sheraton desk, Chippendale chair, Spanish desk, Queen Anne table and a William and Mary dressing mirror. 224pp. 8⅛ x 11¼.
29338-6

THE FOSSIL BOOK: A Record of Prehistoric Life, Patricia V. Rich et al. Profusely illustrated definitive guide covers everything from single-celled organisms and dinosaurs to birds and mammals and the interplay between climate and man. Over 1,500 illustrations. 760pp. 7½ x 10⅛.
29371-8

Paperbound unless otherwise indicated. Available at your book dealer, online at **www.doverpublications.com**, or by writing to Dept. GI, Dover Publications, Inc., 31 East 2nd Street, Mineola, NY 11501. For current price information or for free catalogues (please indicate field of interest), write to Dover Publications or log on to **www.doverpublications.com** and see every Dover book in print. Dover publishes more than 500 books each year on science, elementary and advanced mathematics, biology, music, art, literary history, social sciences, and other areas.